AF871315

Microstructure, Mechanical and Corrosion Properties of Metals, Alloys, and Composite Materials

Guest Editors

Yong Dong
Shichao Liu

MDPI

Basel • Beijing • Wuhan • Barcelona • Belgrade • Novi Sad • Cluj • Manchester

Guest Editors
Yong Dong
School of Materials and Energy
Guangdong University of Technology
Guangzhou
China

Shichao Liu
Institute of Special Solidification and Processing of Advanced Materials
Soochow University
Suzhou
China

Editorial Office
MDPI AG
Grosspeteranlage 5
4052 Basel, Switzerland

This is a reprint of the Special Issue, published open access by the journal *Materials* (ISSN 1996-1944), freely accessible at: https://www.mdpi.com/journal/materials/special_issues/ZJT9985SRZ.

For citation purposes, cite each article independently as indicated on the article page online and as indicated below:

Lastname, A.A.; Lastname, B.B. Article Title. *Journal Name* **Year**, *Volume Number*, Page Range.

ISBN 978-3-7258-8263-2 (Hbk)
ISBN 978-3-7258-8264-9 (PDF)
https://doi.org/10.3390/books978-3-7258-8264-9

© 2026 by the authors. Articles in this reprint are Open Access and distributed under the Creative Commons Attribution (CC BY) license. The reprint as a whole is distributed by MDPI under the terms and conditions of the Creative Commons Attribution-NonCommercial-NoDerivs (CC BY-NC-ND) license (https://creativecommons.org/licenses/by-nc-nd/4.0/).

Contents

About the Editors

Yong Dong

Yong Dong is one of the One Hundred Young Talents of Guangdong University of Technology and an Associate Professor in the Department of Materials Forming and Control Engineering, School of Materials and Energy. They are a founding member of the Low-Carbon Metallurgical Engineering Innovation Cross-Disciplinary Research Institute, and have been ranked in the top 2% of global scientists in 2021, 2022, 2023, 2024 and 2025. They teach courses such as Metallography and Heat Treatment, Thermodynamics of Materials, Principles of Metal Plastic Forming, Material Forming Processes and Mold Design. They are engaged in research on advanced metal materials, advanced metal matrix composites, directional solidification, additive manufacturing, hot-dip galvanized aluminum-magnesium coatings, etc. They have presided over and participated in more than 30 national and provincial funds, as well as horizontal projects. More than of their 100 Sci/EI-indexed papers have been published in international journals such as *Mater Sci Eng A*, *J Alloys Compd*, *Mater Des*, and *Intermetallic*, with over 5,000 citations by other scholars. Their single most highly cited paper has achieved over 1,000 citations, and a h-index over 29. They have served as a reviewer for more than 30 scientific journals such as *J. Mater Sci Tech*, *Mater SCI Eng A*, *J alloys Compd*, and *Mater Des*.

Shichao Liu

Shichao Liu is an Associate Professor in the Department of Metallic Materials Engineering, School of Metallic Materials and Advanced Manufacturing, Soochow University. He was awarded a PhD in Materials Processing and Engineering from Dalian University of Technology. He has published over 80 SCI papers in metallurgy/materials journals, such as *Composites Part B*, the *Journal of Materials Science & Technology*, *Materials and Design*, and the *Journal of Alloys and Compounds*. He has presided over more than 10 projects, including the National Natural Science Foundation of China, the Jiangsu Provincial Natural Science Foundation and other industrial projects. He has applied for, or been granted, eight national invention and utility model patents. Dr. Liu has mentored students in many competitions, including the Undergraduate Innovation and Entrepreneurship Training Program, the National College Student Materials Heat Treatment Innovation and Entrepreneurship Competition, and the National College Student Casting Industrial Design Competition. He also has served as a young editorial member for journals such as *China Foundry*, the *Journal of Materials Heat Treatment*, and *Special Casting & Nonferrous Alloys*, and also as a reviewer for more than 20 scientific journals, including *Composites Part B* and the *Journal of Materials Science & Technology*.

Preface

The study of metallic materials, ranging from pure metals to advanced alloys and composite systems, has long been a cornerstone of materials science and engineering. Their performance in real-world applications is governed by a delicate interplay among three critical pillars: microstructure, mechanical behavior, and corrosion resistance. Understanding and optimizing these interdependent properties is essential not only for scientific discovery but also for industrial innovation and structural integrity.

This Reprint, "Microstructure, Mechanical and Corrosion Properties of Metals, Alloys, and Composite Materials", focuses on the synergistic relationships between the internal architecture (microstructure) of metallic materials and their resulting mechanical response and corrosion performance. The scope spans conventional metals and alloys, as well as emerging metallic composites, including particle-reinforced, fiber-reinforced, and laminated systems. Particular attention is given to how processing routes (casting, deformation, additive manufacturing, and heat treatment) tailor microstructural features (grain size, phase distribution, precipitates, crystallographic texture, and interfaces), and how these features dictate strength, ductility, fatigue, hardness, and susceptibility to various forms of corrosion (uniform, pitting, intergranular, stress corrosion cracking). The aim is not merely to compile our existing knowledge, but to provide a coherent framework that links microstructural design to practical property optimization. The purpose is twofold: first, to advance fundamental understanding of structure–property–performance relationships; second, to serve as a practical reference for predicting and improving material durability and mechanical reliability in service environments.

This Reprint is intended for a broad audience within materials science, metallurgy, mechanical engineering, chemical engineering, and corrosion engineering. It addresses graduate students seeking a comprehensive introduction to microstructure–property relationships; and early-career researchers looking for state-of-the-art insights and critical literature reviews, where metal and composite performance must be reliably predicted and enhanced. Additionally, faculty members may find the Reprint valuable for designing advanced courses on physical metallurgy, mechanical behavior of materials, or corrosion science. Practicing engineers and failure analysts will also benefit from the practical guidelines on material selection and degradation control.

We hope that "Microstructure, Mechanical and Corrosion Properties of Metals, Alloys, and Composite Materials" will serve not only as a repository of knowledge, but also as a source of inspiration for future innovations in durable, high-performance metallic materials.

Yong Dong and Shichao Liu
Guest Editors

materials

MDPI

Communication

Spontaneous Infiltration Behavior of Al Matrix into Carbon Fiber Bundles Induced by Al-Ni Binary Reaction

Jiaming Liu [1,2], Xi Yang [3], Shichao Liu [4,*], Junjia Zhang [5], Yubo Zhang [3,*] and Tingju Li [3]

1 State Key Laboratory of Advanced Equipment and Technology for Metal Forming, Shandong University, Jinan 250061, China; liu.jm@sdu.edu.cn
2 Key Laboratory for Liquid-Solid Structural Evolution and Processing of Materials (Ministry of Education), Shandong University, Jinan 250061, China
3 Key Laboratory of Solidification Control and Digital Preparation Technology (Liaoning Province), Dalian University of Technology, Dalian 116024, China; yangxi@mail.dlut.edu.cn (X.Y.); tjuli@dlut.edu.cn (T.L.)
4 School of Iron and Steel, Soochow University, Suzhou 215006, China
5 Key Laboratory of Lightweight Structural Materials (Liaoning Province), Northeastern University, Shenyang 110819, China; zhangjj@mail.neu.edu.cn
* Correspondence: sc_liu@suda.edu.cn (S.L.); ybzhang@dlut.edu.cn (Y.Z.)

Abstract: In this study, a Ni-coated carbon fiber reinforced Al-matrix (Ni-CF/Al) composite is prepared utilizing a pressure-free infiltration process. The CFs are coated with a layer of Ni through an electroless plating process, which facilitates the spontaneous infiltration behavior driven by the Al-Ni binary reaction. The spontaneous infiltration process, observed via synchrotron radiation in the direction opposite to gravity, demonstrates a fastest velocity of 31.02 ± 1.08 μm/s. By increasing the infiltration temperature, the interfacial microstructure of the composite can be enhanced, characterized by a reduction in un-infiltrated defects and promoted by the interfacial Al-Ni reaction. Notably, large-size Al-Ni intermetallic compounds (IMCs) at the interface are replaced by fine ($Al+Al_3Ni$) eutectic structure, given an optimal fabrication temperature of 720 °C. This contributes to a significantly enhanced ultimate tensile strength (UTS) of the composite, reaching a maximum of 135 ± 4 MPa, which is 159.6% higher than that of the matrix.

Keywords: CF/Al composite; surface metallization; spontaneous infiltration behavior; synchrotron X-ray radiography; interfacial microstructure evolution

1. Introduction

Carbon fiber reinforced Al-matrix (CF/Al) composites have attracted wide attention in transportation and aviation industries due to their excellent properties, including low density, high specific strength, and low thermal expansion coefficient. However, the fabrication of these composite remains challenging [1–4]. For instance, the infiltration of Al into CF bundles cannot spontaneously occur due to the poor interfacial wettability [5].

Applying metal coatings, such as Cu and Ni, to the surface of the CFs is an effective strategy to enhance wettability [6,7]. Previous studies have primarily explored the CF/Al interfacial wettability through titration tests using Al melt and graphite [8]. However, these methods are limited to observing changes in contact angle, and in situ infiltration observation has been rarely studied. In our previous work, the infiltration of Al melt into Ni-coated CFs was observed by synchrotron X-ray radiography [9]. Despite this progress, the mechanism governing spontaneous infiltration under gravitational effects is still unclear.

The optimization of fabrication methods has also been developed to address the aforementioned challenges. It had been demonstrated that at temperatures exceeding 1000 °C, the wettability between Al melt and CFs was improved, with contact angles decreasing to below 90° [10,11]. On this basis, a high-temperature fabrication process has been employed, while the harmful Al_4C_3 phase would be formed at the Al/C interface under these conditions [11,12]. High-pressure conditions were also applied to fabricate CF/Al composites at relatively low temperatures [13,14]. However, the applied external pressure might make it challenging to regulate the interfacial strength. Moreover, typical brittle fracture behavior of the composite would occur due to excessive interfacial strength, adversely affecting the composite performance [15]. Therefore, it is critical to develop a fabrication method under relatively low temperatures and pressures.

In this study, the infiltration process of Al melt into Ni-coated CFs in the anti-gravity direction was innovatively explored using synchrotron X-ray radiography. The spontaneous infiltration behavior, driven by the Al-Ni interfacial reaction, was revealed visually. On this basis, CF/Al composites were successfully fabricated through a pressure-free infiltration process at various temperatures.

2. Experimental

Pure Al and unidirectional CF bundles (T300-3k, Toray Industries Inc., Tokyo, Japan) were utilized as the matrix and reinforcement, respectively. An electroless plating method was used to coat Ni onto the CF surfaces (Figure 1a). The specific details of the deposition process during electroless plating are selected based on our previous study [9]. To observe the infiltration process of molten Al in the anti-gravity direction in vacuum, in situ synchrotron radiation was performed using an X-ray beam with 25 keV energy and a charge-coupled device (CCD) camera (Figure 1b). The Ni-coated CF bundles were arranged atop the Al sheet and tightly connected with it. The infiltration couple was heated to 750 °C at a rate of 10.0 ± 0.2 °C/min and held for 10 min to ensure sufficient infiltration, according to our previous study [16]. Based on the synchrotron radiation experiments, Ni-CF/Al composites were prepared using a pressure-free infiltration method (Figure 1c). Molten Al was poured into a graphite mold containing the fixed CF bundles, followed by infiltration processes in a heat-treatment furnace at various temperatures determined through several experimental attempts (690 °C, 720 °C, and 750 °C) for 6 min that were deemed to be optimal [16]. Additionally, CF/Al composites were prepared in parallel at 750 °C.

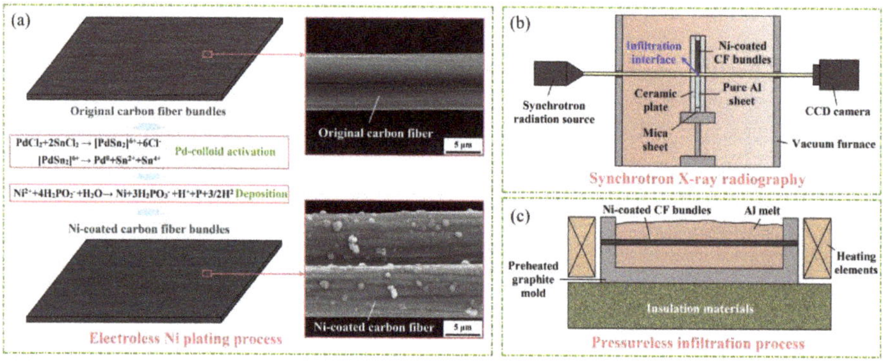

Figure 1. Schematic diagram of the (**a**) electroless Ni plating, (**b**) synchrotron radiation, and (**c**) fabrication process of the CF/Al composite.

The surface morphology of the CFs was observed using an IT800-SHL scanning electron microscope (SEM) (Tokyo, Japan). The microstructure investigation of the composites was performed by SEM, JXA-8530 electron probe microanalysis (EPMA) (Tokyo, Japan), and FEI Talos·F200X transmission electron microscopy (TEM) (Waltham, MA, USA). The density of the composites was measured based on Archimedes principle, with each sample configuration measured five times to ensure accuracy. The sizes of the IMCs were statistically analyzed, with 20 measurements taken for each sample to ensure data reliability. The tensile strength of the composites with a diameter of 5.0 ± 0.1 mm and a gauge length of 15.0 ± 0.1 mm was studied using a UTM 5105 universal testing machine (Kyoto, Japan) at a rate of 0.50 ± 0.02 mm/min. Each configuration of the composite sample was tested 3 times. The volume fraction of the CFs in the composite was approximately 9.0 ± 0.3%.

3. Results and Discussion

Figure 2 presents the synchrotron radiation images of the Ni-coated CF/Al infiltration couple. The initial interface was relatively straight and clearly visible at t_0 (627.0 ± 0.2 °C), indicating that no infiltration for the Al melt had yet occurred (see Figure 2a). At t_0 + 96 s (643.0 ± 0.2 °C), the clearly visible interface disappeared, while a fuzzy region of Al could be observed and began to move upward, which was deemed as the infiltration front (Figure 2b). At this moment, the achieved temperature exceeding the Al-Ni eutectic temperature resulted in the occurrence of the eutectic reaction, even though the Al had not fully melted. At t_0 + 205 s (661.0 ± 0.2 °C), the infiltration front adopted an undulating shape and had moved significantly upward, away from the initial position in the direction opposite to gravity, demonstrating a spontaneous infiltration process due to significantly promoted CF/Al interfacial wettability (Figure 2c). This phenomenon can be attributed to the positive capillary pressure (P_c) caused by the low contact angle between the Al melt and Ni coatings, which was lower than 90° [8], as described by Equations (1) and (2) [17–19]:

$$P_c = \frac{2\gamma \cos\theta}{d} \quad (1)$$

$$h = [\frac{2F(Pc - Pi)}{\eta}]^{1/2} t^{1/2} \quad (2)$$

where γ, θ, and d are the surface tension, contact angle, and the equivalent capillary radius, respectively, and h is the infiltration distance. P_i is the passage resistance. F and η are the osmotic coefficient and coefficient of viscosity, respectively.

As the infiltration process completed at t_0 + 270 s (672.0 ± 0.2 °C), the infiltration fronts exhibited minimal changes (Figure 2d). Three randomly selected points were used to calculate the infiltration velocity and distance, see Figure 2d. The infiltration distance versus time is shown in Figure 2e, with t_0 + 96 s defined as the starting point (0 s). Based on the variation in infiltration velocity, the infiltration process can be then divided into three distinct stages: slow, rapid, and stable infiltration. In the initial slow infiltration stage, the relatively poor interfacial reaction below the melting point temperature resulted in a low infiltration velocity of 1.59 ± 0.06 μm/s. Subsequently, the rapid infiltration stage was observed between 70.0 ± 0.4 s and 130 ± 0.5 s (t_0 + 166.0 ± 0.4 s~t_0 + 226.0 ± 0.5 s), during which the average infiltration velocity significantly increased to approximately 31.02 ± 1.08 μm/s. In the final stable infiltration stage, the velocity gradually decreased to nearly 0 μm/s. Simultaneously, the infiltration distances of the undulating front were quantitatively calculated (Figure 2e), revealing a maximum infiltration distance of approximately 2020 ± 10 μm, with an average value of 1978 ± 27 μm. These results demonstrate that the application of Ni coatings enables the spontaneous infiltration of molten Al into CF bundles, driven by the Al-Ni binary reaction.

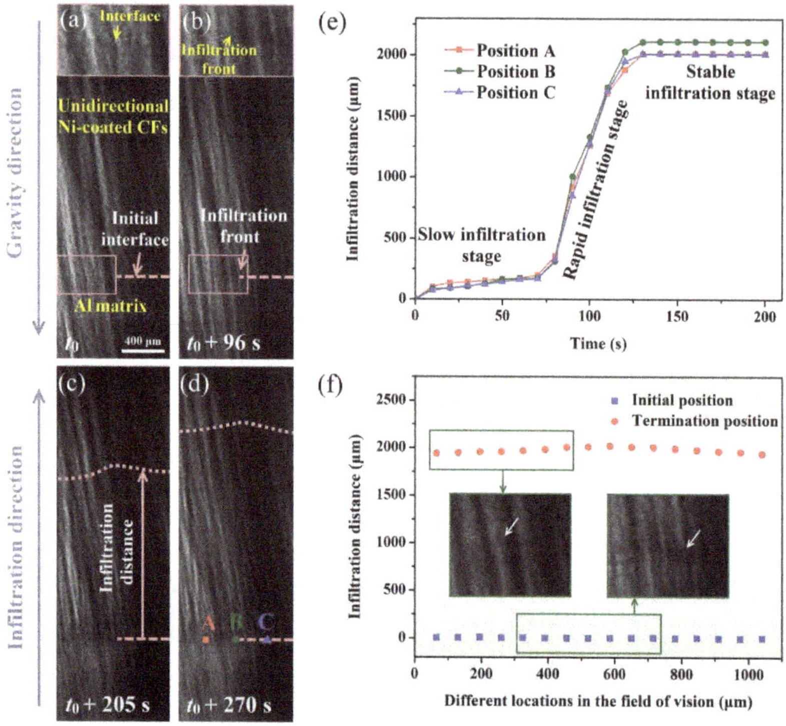

Figure 2. Time-evolved infiltration images for the Ni-coated CFs/Al couple in the direction opposite to gravity: (**a**) t_0; (**b**) t_0 + 96 s; (**c**) t_0 + 205 s; (**d**) t_0 + 270 s; (**e**) infiltration distance versus time curves at various positions; (**f**) initial and termination positions of the infiltration front.

Figure 3a displays the microstructure of CF/Al composites without Ni coatings on the CFs. Due to the poor interfacial wettability, molten Al cannot spontaneously infiltrate into the CF bundles, resulting in the generation of considerable un-infiltrated defects within the CF bundles. Figure 3c–h shows the microstructure of Ni-CF/Al composites fabricated at various infiltration temperatures. At 690 °C, the infiltration of Al melt was significantly promoted, resulting in much fewer defects within the CF bundles, as shown in Figure 3c,d. Simultaneously, larger-size Al-Ni IMCs formed primarily near CFs due to interfacial Al-Ni reactions. The element composition of these IMCs was 74.6 ± 1.2 at.% Al and 25.4 ± 1.2 at.% Ni (typical IMC is shown at Point A), which was determined to be Al_3Ni. Additionally, fine Al_3Ni with eutectic structure was also found [20]. As the temperature increased to 720 °C, the large-sized Al_3Ni phases disappeared and totally transformed to fine eutectic ($Al+Al_3Ni$) phases. This transformation was attributed to further enhanced Al-Ni interfacial diffusion (Figure 3e,f). The Al_3Ni phase was further confirmed by HR-TEM images at the IMC/Al interface and corresponding electron diffraction patterns (Figure 3e). As temperature increased to 750 °C, the fine Al_3Ni phases were gradually reduced owing to the rapid diffusion of Ni (Figure 3g,h). Notably, apparent white impurities could be found within the CF bundles, likely owing to the difficulty for impurities to be removed from CF bundles. The composition of the white impurities was 42.6 ± 2.3 at.% Al and 57.4 ± 2.3 at.% O according to the EDS analysis results (typical impurity is shown at Point B), identifying the impurity as Al_2O_3.

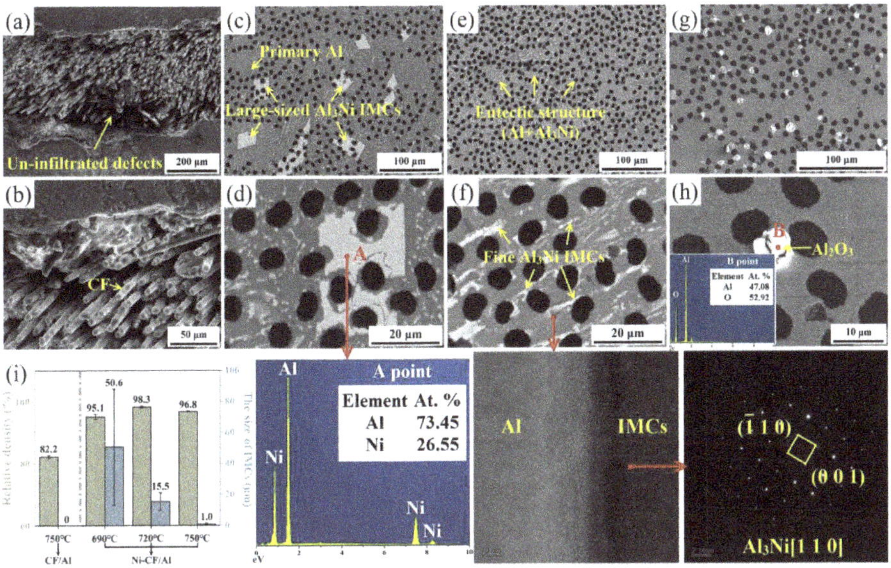

Figure 3. Micromorphologies of (**a**,**b**) CF/Al composites prepared at 750 °C and Ni-CF/Al composites prepared at (**c**,**d**) 690 °C, (**e**,**f**) 720 °C, and (**g**,**h**) 750 °C; (**i**) the relative density of the composites fabricated at varying temperatures and the size of IMCs.

Subsequently, the relative density of the composites fabricated at varying temperatures and the size of the IMCs were measured and are summarized in Figure 3i. Among them, the green columns and labels in Figure 3i represent the relative density of the composites, while the blue columns and labels correspond to the size of the IMCs. By the introduce of Ni coatings onto the CF surfaces, the relative density of the composites significantly increased to 95.1 ± 0.8%, 98.3 ± 0.3%, and 96.8 ± 0.1% at 690 °C, 720 °C, and 750 °C, respectively, compared to that of the composites with desized CFs (82.2 ± 0.4%). Furthermore, the relatively inhomogeneous IMCs with a larger size of 50.6 ± 37.6 μm was noted for the Ni-CF/Al composites fabricated at 690 °C, while gradually decreased size of the IMCs was obtained as the fabrication temperature increased to 720 °C (15.5 ± 5.6 μm) and 750 °C (1.0 ± 0.4 μm). These findings were consistent with the microstructure evolution results described earlier.

The element distribution within the CF/Al composite was further investigated, as shown in Figure 4. With the increased infiltration temperature from 690 °C to 750 °C, the Al-Ni diffusion could be continuously promoted during the pressure-free infiltration process of the composite. Specifically, the uniformity of Ni element distribution was significantly improved with the increasing infiltration temperature. Accordingly, this resulted in a transformation from large-sized Al-Ni IMCs to a fine eutectic structure (Al+Al_3Ni), aligning well with the aforementioned microstructure analysis results at the CF/Al interface regions. The interfacial microstructure evolution can be attributed to the diffusion behavior of Ni into Al melt. It had been demonstrated that the Al-Ni interfacial diffusion could be promoted with the increase in fabrication temperature [21]. In such a case, the diffusion distance of Ni into Al melt would be improved, leading to a relatively lower localized Ni content. This phenomenon, as per the Al-Ni binary phase diagram [22], accounts for the morphological evolution of Al-Ni IMCs and the improved homogeneity in their distribution. In particular, the large-size primary Al_3Ni at the interface changed to a fine (Al+Al_3Ni) eutectic structure.

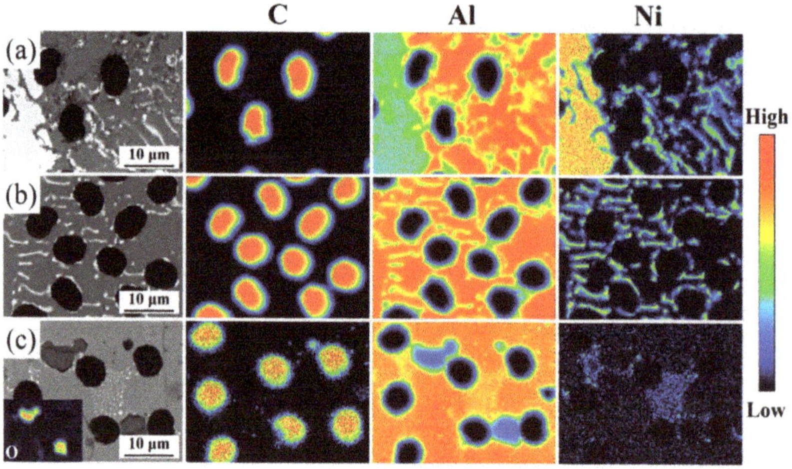

Figure 4. Element distribution in the composites prepared at (**a**) 690 °C, (**b**) 720 °C, and (**c**) 750 °C.

Figure 5a shows the UTS of the composites. The UTS of the CF/Al composites was measured at 60 ± 2 MPa. It increased to 107 ± 1 MPa, 135 ± 4 MPa, and 82 ± 1 MPa for the Ni-CF/Al composites with the increase in infiltration temperature, respectively, which was 105.8%, 159.6%, and 57.7% higher than the Al matrix (52 ± 2 MPa). In this case, the infiltration temperature of 720 °C was identified as optimal. The fracture morphology of composites and a schematic diagram of crack propagation at the interface are shown in Figure 5b–g. Notably, the interfacial failure behavior would significantly affect the composite strength [23]. For composites fabricated at 690 °C, the brittle fracture of large-sized Al_3Ni IMCs was observed, accompanied by the fibers cutting directly (Figure 5b). This phenomenon is attributed to the formation of large-sized Al_3Ni IMCs, which are prone to initiating crack sources and leading to rapid failure of fibers (Figure 5e) [24]. This resulted in a larger multiple fracture number (k) of fibers, which was proved to lower the harmful energy absorption and negatively affect the strength of the composite, according to Equation (3) [25].

$$\sigma_c = \min[(\frac{L_0}{2(\ln 2)e\beta\lambda k})^{\frac{1}{\beta}}\gamma, (\frac{2L_0\tau_b}{e(\beta+1)R_f k})^{\frac{1}{\beta+1}}\gamma^{\frac{\beta}{\beta+1}}] \quad (3)$$

where σ_c is the strength of the composite. L_0 is the reference length, e is the natural constant, and γ and β are the scale and shape parameters, respectively. λ is the ineffective length factor, and k is the multiple fracture number. τ_b is the strength of the interface, and R_f is the fiber radius.

For the case of 720 °C, the fracture morphology exhibited phenomena such as fiber pull-out and fiber cutting directly, accompanied by the presence of more dimples, as shown in Figure 5c. In this scenario, tiny cracks can be deflected by the eutectic structure ($Al+Al_3Ni$), leading to a continuously changed crack propagation path and more energy absorption (see Figure 5f) [26]. Consequently, effective load transferred to the CFs can be obtained, contributing the highest composite strength. As the temperature further increased to 750 °C, the primary failure mechanisms shifted to fiber cutting directly and broken Al_2O_3 (Figure 5d). Notably, the Al_2O_3 impurities acted as crack sources, resulting in rapid crack propagation across the fibers and relatively poor load transfer (Figure 5g). In this case, the strength of the composite instead decreased.

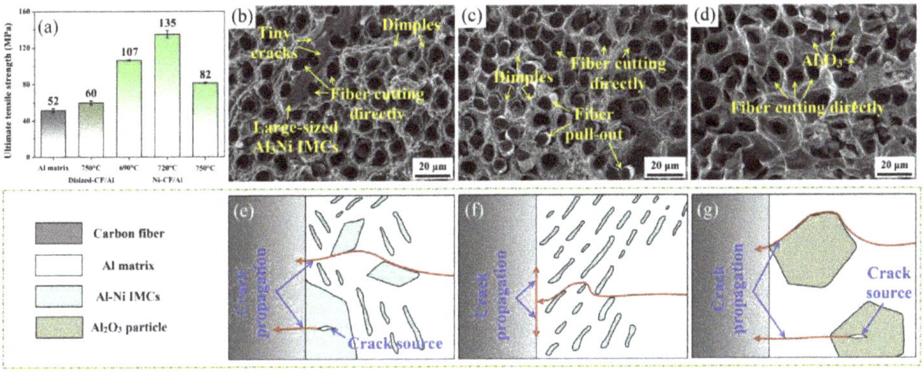

Figure 5. (**a**) Average UTS of CF/Al composites; fracture morphologies of Ni-CF/Al composites prepared at (**b**) 690 °C, (**c**) 720 °C, and (**d**) 750 °C; (**e–g**) corresponding schematic diagram of the crack propagation.

Table 1 presents a comparison of preparation temperature, pressure, and the improvement in UTS relative to the matrix for the composite in this work and those reported in the literature. It can be found that achieving a significant improvement in UTS compared to the matrix typically requires the application of relatively high pressure. In contrast, this study demonstrates a significant improvement in the UTS of the CF/Al composite under pressure-free conditions, even with a much lower CF volume fraction. This eliminates the need for additional pressure during composite preparation, thereby reducing the cost of the preparation equipment. Moreover, the relative simplicity of the process allows for improved efficiency of the composite preparation. The aforementioned factors lead to a potential decrease in the cost of obtained composite parts. Additionally, it was also noticed that for most studies reported in the literature, there were no apparent interfacial IMCs generated at the CF/matrix interface. Despite the localized Al-rich IMCs identified in some research, they have not been further identified or regulated. In the present study, effective Al-Ni IMC regulation and morphology optimization were achieved, facilitating a considerable improvement in the mechanical properties of the CF/Al composite.

Table 1. Comparison of preparation temperature, pressure, and improvement in UTS compared to the matrix for the composite in this work and those reported in the literature.

Ref.	CF	CF Volume Fraction/%	Matrix	Preparation Method	Temperature/°C	Pressure/MPa	Interfacial IMCs	Improvement in UTS/MPa	Percentage of UTS Improvement/%
[14]	NA	20	2024Al	Squeeze casting	750	50	Al-Ni IMCs	168	66.4
[27]	T700	45	AZ91D	Squeeze casting	610	30	/	103	68.7
[28]	T300	NA	A-5Cu-0.5Mn	Plasma sintering	480	50	/	34	20.2
[29]	NA	2	Al-17Si	Liquid metallurgy	710	/	/	14	9.7
This study	T300	9	Al	Pressure-free infiltration	720	/	Al_3Ni	83	159.6

4. Conclusions

In this study, the infiltration behavior of molten Al into Ni-coated CF bundles was in situ investigated in the direction opposite to gravity. Spontaneous infiltration behavior of the Al melt could be obtained, driven by the Al-Ni binary reaction. The largest infiltration velocity was measured at 31.02 ± 1.08 μm/s, with an average infiltration distance of 1978 ± 27 μm. On this basis, the Ni-CF/Al composite was successfully fabricated using a pressure-free infiltration process. With the increase in infiltration temperature, Ni diffusion could be improved, leading to the elimination of defects and a transformation in the morphology of Al-Ni IMCs from large-size primary Al_3Ni to fine ($Al+Al_3Ni$) eutectic structure. This resulted in an effective load transfer and crack propagation deflection, contributing to a significantly enhanced UTS of the Ni-CF/Al composite (135 ± 4 MPa) at the optimal infiltration temperature of 720 °C, representing a 159.6% increase compared to the Al matrix. These findings proved a promising pressure-free fabrication route for fiber/metal composites with enhanced interfacial and mechanical performance.

Author Contributions: J.L.: writing—original draft, conceptualization, investigation, data curation; X.Y.: literature search and validation; S.L.: conceptualization and formal analysis; J.Z.: writing—review and editing, and investigation; Y.Z.: supervision and visualization; T.L.: conceptualization. All authors have read and agreed to the published version of the manuscript.

Funding: This work was financially supported by the National Natural Science Foundation of China [No. 52301054, 52101171, and 51971048] and the China Post-Doctoral Science Foundation [No. 2023M732038].

Institutional Review Board Statement: Not applicable.

Informed Consent Statement: Not applicable.

Data Availability Statement: The original contributions presented in this study are included in the article. Further inquiries can be directed to the corresponding authors.

Conflicts of Interest: We declare that we have no financial and personal relationships with other people or organizations that can inappropriately influence our work; there is no professional or other personal interest of any nature or kind in any product, service, and/or company that could be construed as influencing the position presented in, or the review of, this manuscript.

References

1. Zhang, Y.; Wang, S.; Zhao, X.; Wang, F.; Wu, G. In Situ Study on Fracture Behavior of Z-Pinned Carbon Fiber-Reinforced Aluminum Matrix Composite via Scanning Electron Microscope (SEM). *Materials* **2019**, *12*, 1941. [CrossRef] [PubMed]
2. Wang, Z.; Yang, S.; Du, Z.; Jiang, W.; Zhang, A.; Cai, C.; Yang, W. Micromechanical Modeling of Damage Evolution and Mechanical Behaviors of CF/Al Composites under Transverse and Longitudinal Tensile Loadings. *Materials* **2019**, *12*, 3133. [CrossRef] [PubMed]
3. Wang, S.; Zhang, Y.; Sun, P.; Cui, Y.; Wu, G. Microstructure and Flexural Properties of Z-Pinned Carbon Fiber-Reinforced Aluminum Matrix Composites. *Materials* **2019**, *12*, 174. [CrossRef] [PubMed]
4. Guan, T.; Lu, L.; Chen, Z.; Yang, L. Temperature Resistance Properties of Unidirectional Laminated Cf/SiC-Al Prepared by PIP and Vacuum Pressure Infiltration. *Materials* **2023**, *16*, 5445. [CrossRef]
5. Zhong, K.; Zhou, J.; Zhao, C.; Yun, K.; Qi, L. The effect of nickel coating on the mechanical properties and failure modes of continuous carbon fiber reinforced aluminum matrix composites. *J. Alloys Compd.* **2022**, *904*, 164134. [CrossRef]
6. Oh, S.-I.; Lim, J.-Y.; Kim, Y.-C.; Yoon, J.; Kim, G.-H.; Lee, J.; Sung, Y.-M.; Han, J.-H. Fabrication of carbon nanofiber reinforced aluminum alloy nanocomposites by a liquid process. *J. Alloys Compd.* **2012**, *542*, 111–117. [CrossRef]
7. Mu, W.; Lin, J.; Liu, E.; Zhou, C.; He, W. Fabrication of continuous carbon-fibers reinforced aluminum matrix composites and coating evolution during heat treatment. *J. Mater. Res. Technol.* **2022**, *17*, 1852–1867. [CrossRef]
8. Ip, S.; Sridhar, R.; Toguri, J.; Stephenson, T.; Warner, A. Wettability of nickel coated graphite by aluminum. *Mater. Sci. Eng. A* **1998**, *244*, 31–38. [CrossRef]
9. Liu, J.; Zhang, Y.; Guo, Z.; Liu, S.; Zhang, J.; Guo, E.; Jie, J.; Wang, T.; Li, T. In situ infiltration behavior and interfacial microstructure evolution between Al and carbon fibers. *J. Alloys Compd.* **2021**, *853*, 157120. [CrossRef]
10. Eustathopoulos, N.; Joud, J.; Desre, P.; Hicter, J. The wetting of carbon by aluminium and aluminium alloys. *J. Mater. Sci.* **1974**, *9*, 1233–1242. [CrossRef]
11. Guo, B.; Chen, B.; Zhang, X.; Cen, X.; Wang, X.; Song, M.; Ni, S.; Yi, J.; Shen, T.; Du, Y. Exploring the size effects of Al_4C_3 on the mechanical properties and thermal behaviors of Al-based composites reinforced by SiC and carbon nanotubes. *Carbon* **2018**, *135*, 224–235. [CrossRef]
12. Li, S.-H.; Chao, C.-G. Effects of carbon fiber/Al interface on mechanical properties of carbon-fiber-reinforced aluminum-matrix composites. *Metall. Mater. Trans. A* **2004**, *35*, 2153–2160. [CrossRef]

13. Seong, H.; Lopez, H.; Robertson, D.; Rohatgi, P. Interface structure in carbon and graphite fiber reinforced 2014 aluminum alloy processed with active fiber cooling. *Mater. Sci. Eng. A* **2008**, *487*, 201–209. [CrossRef]
14. Hajjari, E.; Divandari, M.; Mirhabibi, A. The effect of applied pressure on fracture surface and tensile properties of nickel coated continuous carbon fiber reinforced aluminum composites fabricated by squeeze casting. *Mater. Des.* **2010**, *31*, 2381–2386. [CrossRef]
15. Pei, Z.; Li, K.; Gong, J.; Shi, N.; Elangovan, E.; Sun, C. Micro-structural and tensile strength analyses on the magnesium matrix composites reinforced with coated carbon fiber. *J. Mater. Sci.* **2009**, *44*, 4124–4131. [CrossRef]
16. Yang, X.; Zhang, Y.; Huang, J.; Liu, J.; Chen, J.; Li, T. Interfacial microstructure evolution and mechanical properties of carbon fiber reinforced Al-matrix composites fabricated by a pressureless infiltration process. *Mater. Sci. Eng. A* **2024**, *891*, 145968. [CrossRef]
17. Masoodi, R.; Pillai, K. A general formula for capillary suction-pressure in porous media. *J. Porous Media* **2012**, *15*, 775–783. [CrossRef]
18. Scheidegger, A.E. Statistical Hydrodynamics in Porous Media. *J. Appl. Phys.* **1954**, *25*, 994–1001. [CrossRef]
19. Zhang, J.; Liu, S.; Liu, J.; Lu, Y.; Li, T. Electroless nickel plating and spontaneous infiltration behavior of woven carbon fibers. *Mater. Des.* **2019**, *186*, 108301. [CrossRef]
20. Czerwinski, F.; Aniolek, M.; Li, J. Strengthening retention and structural stability of the Al-Al3Ni eutectic at high temperatures. *Scr. Mater.* **2022**, *214*, 114679. [CrossRef]
21. Wang, T.; Cao, F.; Zhou, P.; Kang, H.; Chen, Z.; Fu, Y.; Xiao, T.; Huang, W.; Yuan, Q. Study on diffusion behavior and microstructural evolution of Al/Cu bimetal interface by synchrotron X-ray radiography. *J. Alloys Compd.* **2014**, *616*, 550–555. [CrossRef]
22. Hou, L.; Li, B.; Wu, R.; Cui, L.; Ji, P.; Long, R.; Zhang, J.; Li, X.; Dong, A.; Sun, B. Microstructure and mechanical properties at elevated temperature of Mg-Al-Ni alloys prepared through powder metallurgy. *J. Mater. Sci. Technol.* **2017**, *33*, 947–953. [CrossRef]
23. Chao, X.; Qi, L.; Ma, W.; Ge, J.; Tian, W. Monte carlo based strategy to simulate the microstructure evolution of the short-fiber reinforced metal matrix composites. *Mater. Today Commun.* **2022**, *33*, 104275. [CrossRef]
24. Wright, R.; Knibloe, J. The influence of alloying on the microstructure and mechanical properties of P/M Ni_3Al. *Acta Metall. Mater.* **1990**, *38*, 1993–2001. [CrossRef]
25. Shioya, M.; Yasui, S.; Takaku, A. Relation between interfacial shear strength and tensile strength of carbon fiber/resin composite strands. *Compos. Interface* **1999**, *6*, 305–323. [CrossRef]
26. Fan, Y.; Huang, K.; Makhlouf, M.M. Precipitation strengthening in Al-Ni-Mn alloys. *Metall. Mater. Trans. A* **2015**, *46*, 5830–5841. [CrossRef]
27. Zhou, J.; Yuan, M.; Li, Z.; Meng, H.; Zhang, T.; Qi, L. A great improvement of tensile properties of Cf/AZ91D composite through grafting CNTs onto the surface of the carbon fibers. *Mater. Sci. Eng. A* **2019**, *762*, 138061. [CrossRef]
28. Yang, Q.; Liu, J.; Li, S.; Wang, F.; Wu, T. Fabrication and mechanical properties of Cu-coatedwoven carbon fibers reinforced aluminum alloy composite. *Mater. Des.* **2014**, *57*, 442–448. [CrossRef]
29. Ramesh, C.; Prasad, T. Friction and wear behavior of graphite-carbon short fiber reinforced Al–17% Si alloy hybrid composites. *J. Tribol.* **2009**, *131*, 014501. [CrossRef]

Disclaimer/Publisher's Note: The statements, opinions and data contained in all publications are solely those of the individual author(s) and contributor(s) and not of MDPI and/or the editor(s). MDPI and/or the editor(s) disclaim responsibility for any injury to people or property resulting from any ideas, methods, instructions or products referred to in the content.

materials MDPI

Article

Microstructure Evolution and Mechanical Properties of Dual-Phase $AlCrFe_2Ni_2$ High-Entropy Alloy Under High-Strain-Rate Compression

Hang Yan [1], Yu Wang [2], Xilin Gan [1], Yong Dong [1,*], Shichao Liu [3], Shougang Duan [1,4] and Lingbo Mao [1]

[1] Innovation & Interdisciplinary Institute of Low Carbon Metallurgical Engineering, School of Materials and Energy, Guangdong University of Technology, Guangzhou 510006, China; yanhang0317@163.com (H.Y.); 17502008923@163.com (X.G.); sgduan0410@163.com (S.D.); lbmao@gdut.edu.cn (L.M.)
[2] School of Materials Science and Engineering, North University of China, Taiyuan 030051, China; wangyu@nuc.edu.cn
[3] School of Iron and Steel, Soochow University, Suzhou 215021, China; sc_liu@suda.edu.cn
[4] School of Materials Science and Engineering, South China University of Technology, Guangzhou 510640, China
* Correspondence: dongyong5205@163.com

Abstract: This paper investigates the effect of strain rate on the mechanical deformation and microstructural development of dual-phase $AlCrFe_2Ni_2$ high-entropy alloy during quasi-static and dynamic compression processes. It is revealed that the as-cast $AlCrFe_2Ni_2$ alloy is composed of a mixture of FCC, disordered BCC, and ordered B2 crystal structure phases. The alloy shows excellent compressive properties under quasi-static and dynamic deformation. The yield strength exceeds 600 MPa while the compressive strength is more than 3000 MPa at the compression rates of 30% under quasi-static conditions. Under dynamic compression conditions, the ultimate compression stresses are 1522 MPa, 1816 MPa, and 1925 MPa with compression strains about 12.8%, 14.7%, and 18.2% at strain rates of 1300 s^{-1}, 1700 s^{-1} and 2100 s^{-1}, respectively. The dynamic yield strength is approximately linear with strain rate within the specified range and exhibit great sensitivity. The strong localized deformation regions (i.e., adiabatic shear bands (ASBs)) appear in dynamically deformed samples by dynamic recrystallization due to the conflicting processes of strain rate hardening and heat softening.

Keywords: high-entropy alloys; dynamic compression; adiabatic shear bands; microstructure

1. Introduction

For conventional alloys, such as titanium alloys, magnesium alloys, and aluminum alloys, they are traditionally designed as one or two principal elements with minor additions of the other elements. Recently, a new alloy system, i.e., high-entropy alloys (HEAs), was designed differently from traditional counterparts. For HEAs, four or more principal elements are present in proportions of 5% to 35%, with the mixing entropy more than 1.5R [1,2]. HEAs usually form a simple solid solution phase [1], for instance, BCC, FCC, and HCP phases. These phases can either exist alone or as a dual-phase coexistence [2]. Based on previous research, single-phase HEAs with a BCC structure exhibit low plasticity properties but high tensile strength, whereas HEAs with a single-phase FCC structure typically have good plasticity but low tensile strength [3–5]. In order to balance the properties of HEAs, novel eutectic high-entropy alloys (EHEAs) consisting of a composite microstructure with FCC and BCC phases were designed and proposed by Lu et al. [6]. Until now,

 https://doi.org/10.3390/ma18061191

great potential has been recognized in many applications for these EHEAs, owing to their excellent fracture toughness [7], good cryogenic properties [7,8], exceptional wear, and fatigue resistances [9,10]. In addition, it is simple to adapt EHEAs with good flowability, castability, and high strength and ductility to large-scale industrial manufacturing [11,12].

Currently, several EHEA systems have been developed, including $AlCoCrFeNi_{2.1}$ [6], $Al_{17}Ni_{34}Ti_{17}V_{32}$ [13], $AlCr_{1.3}TiNi_2$ [14], and $AlCrCuFeNi_2$ [15]. Owing to better practicality and low cost, the $AlCrFe_2Ni_2$ alloy, with similar microstructure and properties to the $AlCoCrFeNi_{2.1}$ alloy, was investigated. It is indicated that the $AlCrFe_2Ni_2$ alloy consists of the FCC and A2/B2 spinodal decomposition phase and presents better fracture strength and ductility [16]. In addition, the dynamic mechanical response of HEAs has also received a lot of attention. To use HEAs in extreme industrial environments, there is an urgent need for a comprehensive understanding of their performance and mechanisms during high-speed compression [17–21].

According to the available literature, in addition to the dislocation-related mechanism, twinning mechanism, grain boundary-related mechanism, diffusion creep mechanism, and dynamic recrystallization mechanism, the shear bands are formed in HEAs at high strain rates or under large deformation conditions [17–21]. The adiabatic shear band (ASB) is a localized region formed by a very large shear deformation in a very short time. In this region, along with the adiabatic temperature rise, the microstructure and properties of the material undergo significant changes, such as grain refinement and an increase in dislocation density. The ASB is an ideal mode of deformation, and the presence of the ASB is conducive to the enhancement of the penetration properties of the material. These traits differ significantly from those that occur when quasi-static characteristics are present. The dynamic behavior of $Al_{0.6}CoCrFeNi$ biphasic (FCC and BCC) alloys was examined by Wang et al. [22]. The findings demonstrated that the yield strength could be clearly increased by increasing the strain rate. Tian et al. found that the AlCoCrFeNi alloy exhibited extremely high strain rate sensitivity by investigating the alloy at different strain rates [23]. The dynamic deformation mechanism of the CrCoFeNi high-entropy alloy was investigated by Cao et al. [24]. It was found that massive dislocation nucleation and drag increased the strain rate sensitivity of the alloy at high strain rates, and the alloy also exhibited significant strain-hardening capability. Yuan et al. prepared $Ni_{32}Co_{30}Cr_{10}Fe_{10}Al_{18}$ EHEAs by laser metal deposition, and it was found that the presence of the FCC phase caused the alloy to exhibit high ductility and strain-hardening ability at high temperatures and high strain rates [25]. Jeong Min Park et al. [26] found that the CoCrFeMnNi alloy showed strong strain rate sensitivity. Compression experiments on the $AlCrCuFeNi_2$ alloy at a strain rate of 10^{-3} s^{-1}~3000 s^{-1} were conducted by Ma et al. [15]. They discovered that over a range of strain rates there was a roughly linear connection between dynamic yield strength and strain rate. N. Kumar et al. [27] conducted compression tests on the $Al_{0.1}CrCoFeNi$ alloy using split-Hopkinson pressure bar (SHPB) and noted notable work hardening and strong strain rate sensitivity, as well as a large number of twinned grains [28].

Herein, in the present study, the primary emphasis is to provide more detailed insights into the compression behavior of the dual-phase $AlCrFe_2Ni_2$ alloy under both static compression and dynamic impact. In this context, the impact of strain rates on the microstructural development and mechanical characteristics of the $AlCrFe_2Ni_2$ alloy is assessed using the SHPB technique. Correspondingly, these fundamental processes for the development of microstructural features and mechanical characteristics are also examined and explained.

2. Experimental

2.1. Alloy Fabrication

For the present study, the experimental $AlCrFe_2Ni_2$ alloys were fabricated by using a vacuum high-frequency induction melting furnace, and all raw elemental metals have purity levels exceeding 99.9 wt.%. Melting and casting were conducted by backfilling with argon (Ar) under vacuum pressure up to 3×10^{-3} Pa at 0.01 atm.

2.2. Property Testing

For the static tests, cylindrical specimens that were about 10 mm tall and 5 mm in diameter were wire-cut and then ground. The deformation behavior of $AlCrFe_2Ni_2$ alloy at the strain rate in the range of 10^{-4} s^{-1}~10^{0} s^{-1} was examined using a universal testing machine (AGS-X Plus100 KN, Japan), three samples being tested at each strain rate.

Following compression tests, samples were sliced along the stress loading direction, and silicon carbide paper with mechanical thinning was used to mechanically polish the surface parallel to the compression direction. EBSD analysis was used for microstructural characterization and orientation imaging microscope analysis. Samples measuring around 8 mm in diameter and 5 mm in height were prepared for dynamic compression. An SHPB with strain rates ranging from 1300 s^{-1} to 2100 s^{-1} was used for dynamic compression experiments at ambient temperature. The force applied to the deformation zone when the cylindrical sample is dynamically loaded by the SHPB was computed using the information gathered by the strain gauges on the incident and transmitted bars. The function of time t was used to calculate the strain rate $\dot{\varepsilon}$, true strain ε and true stress σ. The following is how the equations are expressed: [29]:

$$\dot{\varepsilon} = -\frac{2C_0}{L_0}\varepsilon_r(\mathrm{t}) \tag{1}$$

$$\varepsilon = -\frac{2C_0}{L_0}\int_0^l \varepsilon_r(t)dt \tag{2}$$

$$\sigma = \frac{A_0}{A}\mathrm{E}\varepsilon_t(\mathrm{t}) \tag{3}$$

where E is the elastic modulus of the incident and transmission bars, and A_0. And C_0 are the wave propagation velocities in the bars. A is the cross-section area of the incident and transmission bars, and L_0 is the height and cross-section area of sample. $\varepsilon_r(t)$ and $\varepsilon_t(\mathrm{t})$ are the experimentally measured strains of the incident and transmitted stress pulses on the SHPB. You can find more information about the SHPB and the stress and strain calculations elsewhere.

2.3. Microstructure Characterization

An X-ray diffractometer (XRD, EMPYREAN with Cu radiation target) was used to identify the phase. It has a scanning rate of 4° min^{-1} and a 2θ scanning range of 20° to 100°. Energy dispersive spectrometry (EDS) in conjunction with scanning electron microscopy (SEM, JEOL-JSM-IT800) and transmission electron microscopy (TEM, Talos F200S) were utilized to describe the microstructural evolution and deformation process of the $AlCrFe_2Ni_2$ alloy. Metallographic polishing and twin-jet electropolishing (electrolyte: 90 vol.% ethanol + 10 vol.% perchloric acid) were used to produce the thin-foil sample for TEM analysis. Samples were prepared by mechanically thinning after normal metallographic polishing for electron backscatter diffraction (EBSD) examination.

3. Results and Discussion

3.1. Microstructural Analysis

Figure 1 shows the X-ray diffraction pattern of the as-cast $AlCrFe_2Ni_2$ alloy. The XRD result indicates that the alloy is composed of a mixture of FCC, disordered BCC, and ordered B2 crystal structures. In addition, due to the same crystal structure and similar lattice parameter, the pattern presents the overlapping disordered BCC and B2 Bragg reflection, which was also found in the other HEAs [30].

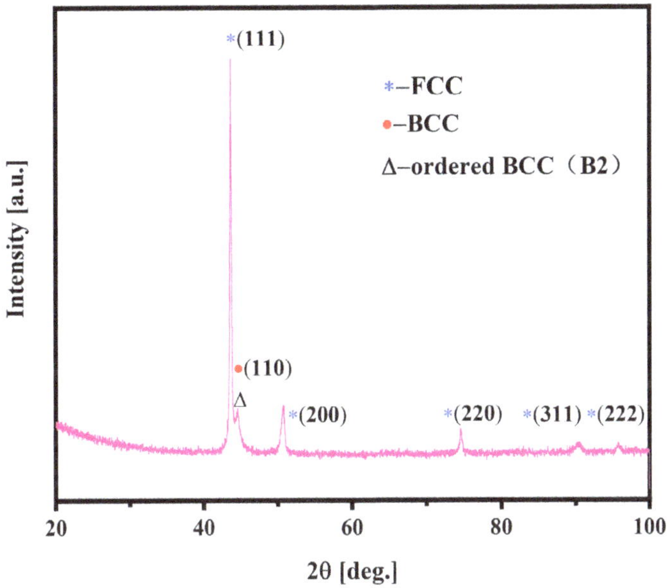

Figure 1. X-ray diffraction of the as-cast $AlCrFe_2Ni_2$ alloy.

Figure 2 shows the SEM images of the as-cast $AlCrFe_2Ni_2$ alloy. From Figure 2a, it can be seen that the alloy mainly consists of noodle-like phases and triangular phases. Further observed by the high-magnified SEM image in Figure 2b, the length and size of the noodle-like phases are not uniform, the other phases are not always triangular in shape, and there are various other shapes. Figure 2c,d shows enlarged images of Figure 2b. The area of the triangle that can be seen in Figure 2c is composed of interconnected light and dark microstructures, which is characteristic of the microstructure of the spinodal decomposition microstructure. The noodle-like phases are clearly visible in Figure 2d, and there are fine dark phases present between the noodle-like phases [16,31,32].

Figure 3 shows the EBSD phase diagram and IPF diagram of the as-cast $AlCrFe_2Ni_2$ alloy. From Figure 3a, it can be seen that the alloy is mainly formed by the FCC phase and BCC/B2 phase, which is consistent with the XRD analysis results, where the FCC phase accounts for 63.7% and the BCC phase accounts for 36.3%. Combined with Figure 2, the noodle-like phase is the FCC phase, and the other phases are the BCC and B2 phases. As can be seen from Figure 3b, the grain orientations of the noodle-like phases are disordered, whereas the orientations of the alternating phases are partly uniform and partly disordered, which suggests that the remaining phases are composed of the BCC phase and the B2 phase.

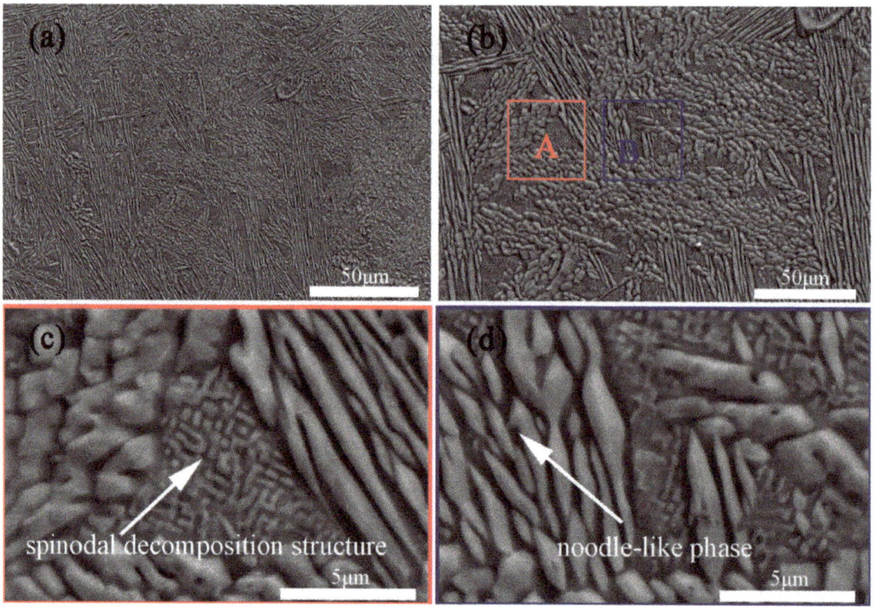

Figure 2. Microstructure of the as-cast $AlCrFe_2Ni_2$ alloy: (**a**) SEM image; (**b**) the high-magnified SEM secondary electron image; (**c**) and (**d**) are magnified images of regions A and B in (**b**), respectively.

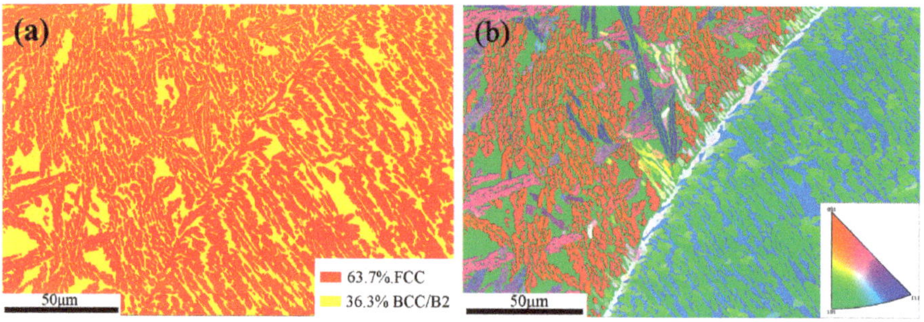

Figure 3. (**a**) EBSD phase diagram of the as-cast $AlCrFe_2Ni_2$ alloy; (**b**) EBSD inverse pole figure (IPF).

Further detailed analysis on the microstructure of the $AlCrFe_2Ni_2$ alloy was carried out by TEM, as seen in Figure 4. Figure 4a,c shows bright-field images of the $AlCrFe_2Ni_2$ alloy. It is evident that the alloy is composed of noodle-like phases and a spinodal decomposition microstructure, which is an alternating mixture of the two phases. Figure 4b,d shows the selected area electron diffraction (SAED) pattern of the alloy $AlCrFe_2Ni_2$. From Figure 4b and the dark-field image (Figure 4c), it can be obtained that the noodle-like phase is the FCC phases, which is also in agreement with the results of the EBSD phase map. From Figure 4d, it can be seen that the modulated structure is a mixture consisting of BCC and B2 phases, and combined with the dark-field image (Figure 4f), in the modulated structure, the matrix has a BCC structure and the particle-shaped phase has a B2 structure. Figure 5b–f shows the element distribution of the $AlCrFe_2Ni_2$ alloy, from which it can be seen that the $AlCrFe_2Ni_2$ alloy forms BCC (Cr, Fe)-rich inter-dendritic matrices, and B2 (Al, Ni)-rich particles (similar to those in Ref. [33]).

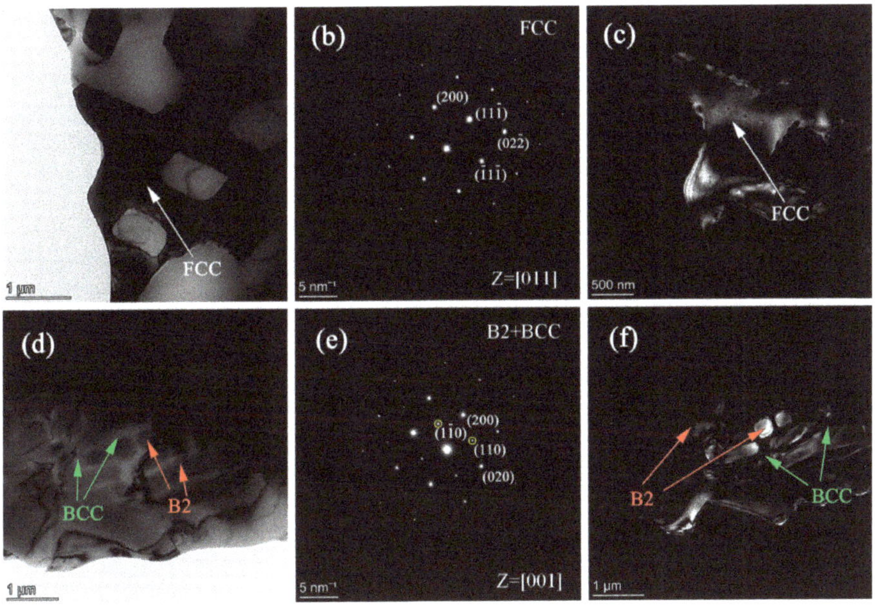

Figure 4. (**a**,**d**) Bright-field images of the $AlCrFe_2Ni_2$; (**b**,**e**) selected area diffractions from zone axes [011] and [001], respectively; (**c**,**f**) dark-field images.

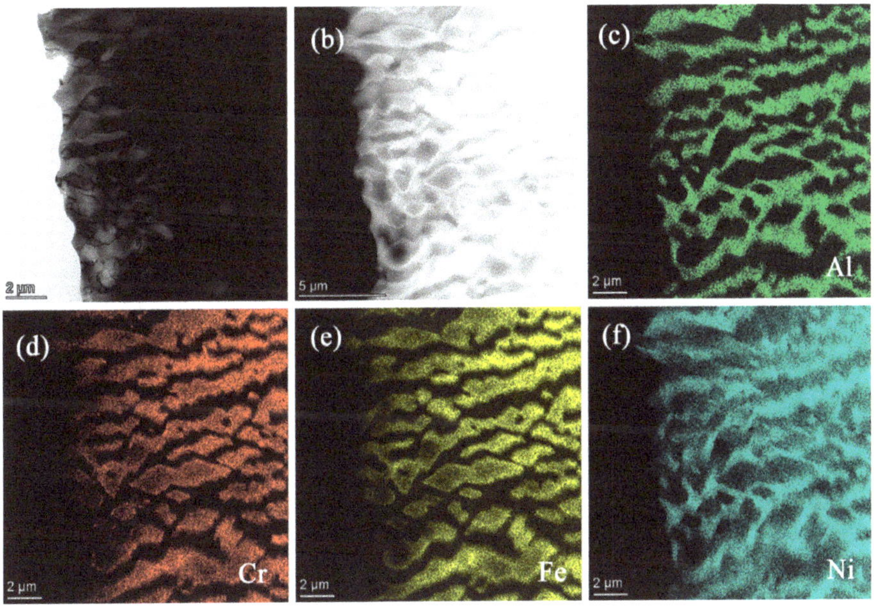

Figure 5. TEM images of the as-cast $AlCrFe_2Ni_2$ alloy: (**a**) bright-field TEM image; (**b**–**f**) the element distribution by TEM-EDS.

3.2. Mechanical Characteristics Under Compression Testing

Figure 6 shows the true stress–strain curve of the $AlCrFe_2Ni_2$ alloy at different strain rates. The true compression stress–strain curves of the $AlCrFe_2Ni_2$ alloy under quasi-static deformation at room temperature are shown in Figure 6a. From Figure 6a, it can be seen that when the strain rate is 10^{-4} s^{-1}–10^{0} s^{-1}, the alloy shows significant work hardening, but there is no significant change in the yield strength value with increasing strain rate. The inset in Figure 6a shows the sample after compression, and the sample did not fracture, which indicates that the $AlCrFe_2Ni_2$ alloy has good plasticity. Thus, the maximum strain is taken to be 30%, and the corresponding ultimate compression stress are more than 3000 MPa. Correspondingly, the compression properties of the $AlCrFe_2Ni_2$ alloy are shown in Table 1. Figure 6b presents the compression performance of the $AlCrFe_2Ni_2$ alloy at high strain rates. It can be seen from Figure 6b that the yield strength changes significantly with the increase in strain rate. Furthermore, a positive strain rate effect is indicated by the dynamic yield stress, and the flow stress shows an upward trend with the rise of strain rate. The compression properties of the $AlCrFe_2Ni_2$ alloy are shown in Table 2. In conjunction with the accompanying graph from Figure 6b, because of the transient loading time, the surface of the $AlCrFe_2Ni_2$ alloy has not changed in any way, the alloy has failed, and therefore severe changes should have occurred within the alloy at high strain rates [34].

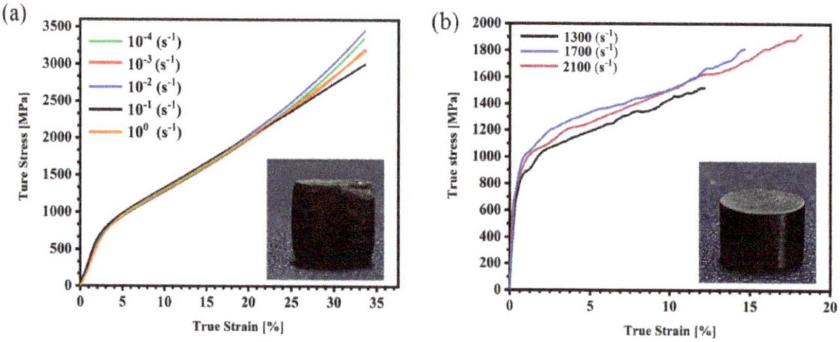

Figure 6. (**a**) Static compression stress–strain curves of the $AlCrFe_2Ni_2$ high-entropy alloy under quasi-statics; (**b**) dynamic compression stress–strain curves at various strain rates.

Table 1. A summary of the compressive properties of the $AlCrFe_2Ni_2$ alloy under quasi-static deformation at room temperature.

Strain Rate [s^{-1}]	Yield Strength [MPa]	Plastic Strain [%]
1×10^{-4}	606.1 ± 3.0	>30.0
1×10^{-3}	616.3 ± 3.1	>30.0
1×10^{-2}	691.6 ± 3.5	>30.0
1×10^{-1}	655.8 ± 3.3	>30.0
1×10^{0}	658.3 ± 3.3	>30.0

Table 2. A summary of the compressive properties of the $AlCrFe_2Ni_2$ alloy under dynamic deformation at room temperature.

Strain Rate [s^{-1}]	Yield Strength [MPa]	Compressive Stress [MPa]	Plastic Strain [%]
1300	891.9 ± 4.5	1521.8 ± 7.6	12.8 ± 2.3
1700	1132.4 ± 5.7	1815.5 ± 9.1	14.7 ± 2.0
2100	1091.7 ± 5.5	1925.3 ± 9.6	18.2 ± 1.6

The strain rate sensitivity (m) is used to describe the dependence of strength/stress on strain rate. It can be defined by the slope of the log strength σ versus log strain rate $\dot{\varepsilon}$ at constant strain ∂:

$$m = \frac{\partial \ln \sigma}{\partial \ln \dot{\varepsilon}} \tag{4}$$

The yield strengths are roughly 606 MPa and 658 MPa at strain rates of 10^{-4} s^{-1} and 10^0 s^{-1}, respectively. The $AlCrFe_2Ni_2$ alloy m under quasi-static deformation is estimated to be around 0.0089. It is worth noting that it is of the same order of magnitude as the conventional metals: Cu (~0.006) and Ni (~0.0028) [34], but an order of magnitude lower than that of partially eutectic high-entropy alloys (~0.028) [35]. The yield strengths are roughly 892 MPa and 1132 MPa at strain rates of 1300 s^{-1} and 2100 s^{-1}, respectively, and the m of the $AlCrFe_2Ni_2$ alloy is estimated to be about 0.42 at high strain rates, which is much higher than the value under quasi-static deformation, and is also consistent with the image pattern of the true stress–strain curve. The present reason for the sudden increase in the strain rate sensitivity coefficient may be due to phonon drag effects [24,36]. Elastic lattice vibrations that go through crystals are called phonons. The phonon drag effect is the result of the interaction between phonons and dislocations, which creates viscous resistance to dislocation sliding. The phonon drag effect is known to be caused by a number of ways [18].

The yield strength for the $AlCrFe_2Ni_2$ alloy is plotted against the logarithmic strain rate in Figure 7. It can be seen from Figure 7 that the trend of yield strength tends to a straight line when the strain rate is 10^{-4} s^{-1}–10^3 s^{-1}, and when the strain rate is greater than 10^3 s^{-1}, the yield strength of the alloy has a clear upward trend with the increase in strain rate. It is evident that under dynamic conditions, the effect of strain rate on yield strength is significantly higher. They are therefore divided into two regions where the strength fluctuates with the strain rate: the area where phonon drag influences dislocation motions ($\dot{\varepsilon} > 10^3$ s^{-1}) and the area of thermally induced dislocation glide ($\dot{\varepsilon} < 10^3$ s^{-1}) [20,27,37,38]. In the region of $\dot{\varepsilon} < 10^3$ s^{-1}, the dislocations engage with the surrounding regions' isotropic phonon flux, and the dislocation velocity is so low that the phonon drag effect barely affects the dislocation motion [18,27]. In the region of $\dot{\varepsilon} > 10^3$ s^{-1}, viscous phonon drag significantly affects dislocation motion [18,27,39]. Fast moving dislocations would cause phonon scattering through the strain field. Moreover, dislocations could absorb phonon energy, which leads to phonon scattering, and the kinematic viscosity of dislocations increases due to the phonon dispersion.

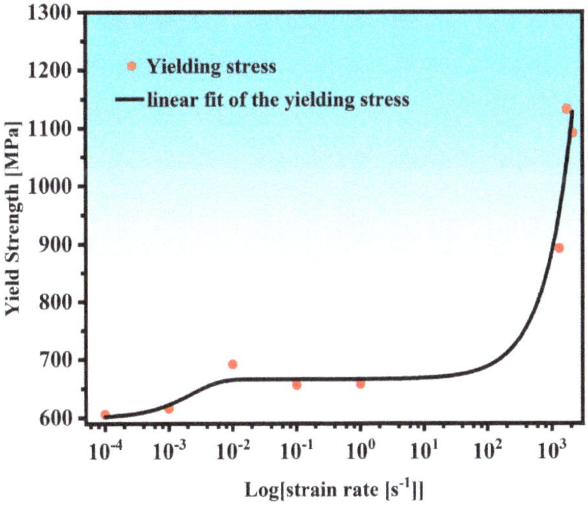

Figure 7. The yield strength as a function of the logarithmic strain rate for the $AlCrFe_2Ni_2$ high-entropy alloy.

3.3. Microstructure Evolution

The IPF and IQ + misorientation border maps (BMs) at a strain rate of 1300 s^{-1} and 2100 s^{-1} are displayed in Figure 8, respectively. From Figure 8a, it can be seen that there is a distinct black area in the center, which is an ASB region, while in Figure 8c, only smaller discontinuous ASB zones can be observed. In addition, a large number of highly refined submicron grains can be seen in the red-boxed portion of Figure 8a,c. In Figure 8c, although many ASB regions are observed, the number of fine grains with HAGBs is significantly less than that in the regions with a strain rate of 1300 s^{-1}. The dynamic recrystallization (DRX) is the cause of the grain refinement observed in the dynamically deformed materials, such as the ASBs frequently seen in a variety of deformation materials, which may be the source of DRX [38,40,41]. The ASB at 1300^{-1} s^{-1} is a narrow region of high-concentration fine grains. These ASBs cause cracks to form, and eventually shear damage, which explains the poor performance of strain rate at 1300^{-1} s^{-1}. The emergence of ASBs may be due to the conflicting processes of strain rate hardening and heat softening [18]. These two processes, strain-hardening and heat-softening effects brought on by dislocation proliferation/stacking or twin evolution, would compete with one another during dynamic deformation. The strain-hardening effect would eventually be overcome by the contribution of thermal softening to the flow stresses, which would result in the production of ASBs. As can be seen from Figure 8b,d, there are a large number of blue regions at high strain rates. The blue region indicates high-angle grain boundaries (HAGBs) with dislocation angles more than 15°, whereas the red and green lines correspond to low-angle grain boundaries (LAGBs) less than 15°. LAGBs provide favorable conditions for dislocation proliferation due to their special structure and higher energy states, which explains the increase in yield strength at high strain rates.

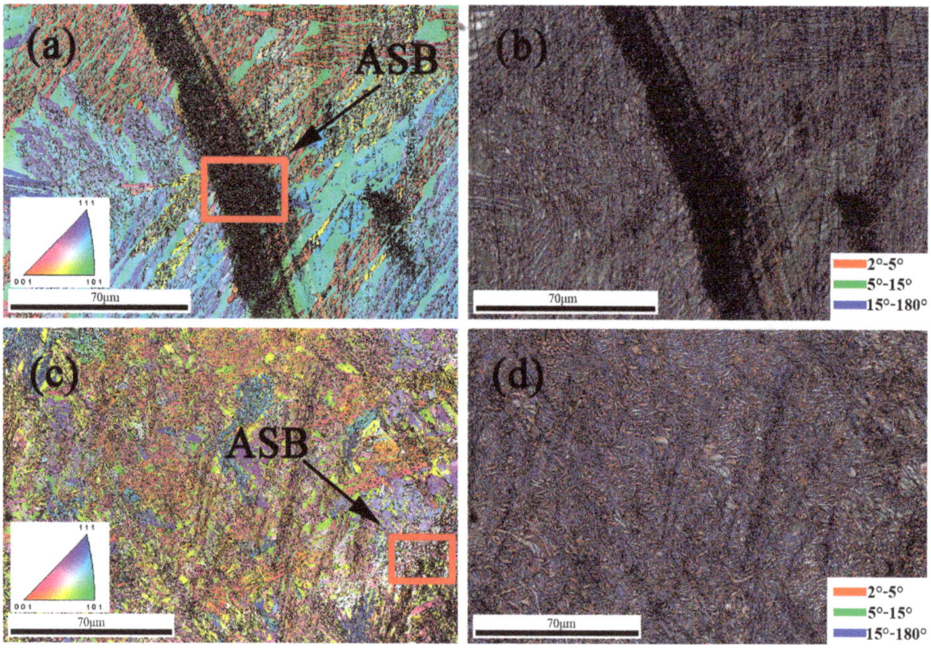

Figure 8. Orientation image maps with the IPF and IQ + misorientation boundary maps of the samples after high-strain compression: (**a**,**b**) at 1300 s^{-1}; (**c**,**d**) at 2100 s^{-1}.

The work-hardening rate (WHR, $d\sigma/d\varepsilon$, where σ is the flow stress and ε is the plastic strain) behavior as a function of the plastic strain at a strain rate of 1300 s^{-1} and 2100 s^{-1} is displayed in Figure 9. It has been recognized that higher strain rates produce a larger WHR, especially in the initial plasticity stage. This characterization can be explained by the expression [42]:

$$\overline{v} = A\tau^{m} \tag{5}$$

where A and m are the constants for specific materials and $\overline{v}$ is the dislocation movement velocity. The initial shear stress for dislocation movement is denoted by τ. The WHR and the strain rates are positively correlated, as seen in Figure 9. The higher strain rates result in the larger shear stress and the higher dislocation motion rate. Two distinct zones can be seen inside the curve. In one region (true plastic strain < 2%), the strain-hardening rate (SHR) drops significantly as the strain increases. In this region, the strain rate is proportional to the work-hardening rate, which is related to the formation of adiabatic shear bands. At a strain rate of 2100 s^{-1}, although there is also the formation of ASBs, the distribution of the shear bands of the fine scattering does not cause the alloy to fail quickly, but rather has a certain strengthening effect. In another region (true plastic strain > 2%), the strain-hardening rate drops less steeply; the plastic deformation in this region is dominated by dislocation slip and the rising dislocation annihilation rate, but the SHR does not rise due to the effect of thermal softening [43].

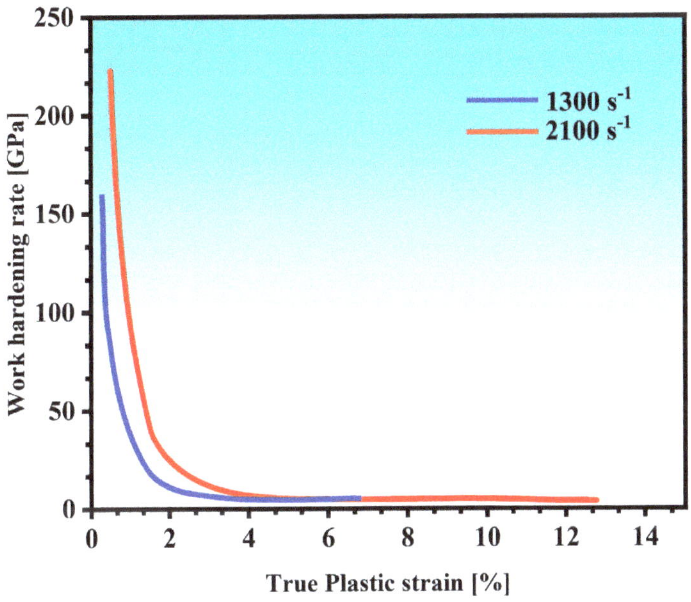

Figure 9. The work-hardening rate (WHR): $d\sigma/d\varepsilon$ (σ is the flow stress and ε is the plastic strain).

4. Conclusions

In this work, the effect of strain rate on the microstructural evolution and compression behavior of the dual-phase $AlCrFe_2Ni_2$ alloy was studied. The main conclusions are as follows:

1. The as-cast $AlCrFe_2Ni_2$ alloy is composed of a mixture of Fe,Ni-enriched FCC phases, disordered Fe,Cr-enriched BCC phases, and ordered Al,Ni-enriched B2 phases.

2. The $AlCrFe_2Ni_2$ alloy shows excellent compressive properties under quasi-static deformation. The yield strength exceeds 600 MPa while the compressive strength is more than 3000 MPa at the compression rates of 30%.
3. The ultimate compression stresses are 1522 MPa, 1816 MPa, and 1925 MPa with compression strains about 12.8%, 14.7%, and 18.2% at strain rates of 1300 s^{-1}, 1700 s^{-1}, and 2100 s^{-1}, respectively. Strain rate sensitivity ranges from 0.0089 at low strain (10^{-4} s^{-1}–10^{0} s^{-1}) to 0.42 at high strain rates (1300 s^{-1}–2100 s^{-1}), which is due to the phonon drag effect at high strains.
4. At high-strain-rate compression, a adiabatic shear phenomenon occurs. The adiabatic shear bands formed by dynamic recrystallization due to the conflicting processes of strain rate hardening and heat softening.

Author Contributions: Methodology, H.Y., Y.W. and S.D.; Validation, Y.W., S.L. and S.D.; Formal analysis, S.D.; Investigation, H.Y., X.G. and S.D.; Resources, Y.D., S.L. and L.M.; Data curation, S.L.; Writing—original draft, H.Y.; Writing—review & editing, X.G. and S.L.; Visualization, Y.W., X.G., S.D. and L.M.; Supervision, L.M.; Project administration, Y.D.; Funding acquisition, Y.D. All authors have read and agreed to the published version of the manuscript.

Funding: This research was supported by the Natural Science Foundation of Guangdong Province (Nos. 2022A0505050052 and 2022A1515012591).

Institutional Review Board Statement: Not applicable.

Informed Consent Statement: Not applicable.

Data Availability Statement: The original contributions presented in this study are included in the article. Further inquiries can be directed to the corresponding author.

Conflicts of Interest: The authors declare that they have no known competing financial interests or personal relationships that could have appeared to influence the work reported in this paper.

References

1. Yeh, J.W.; Chen, S.K.; Lin, S.J.; Gan, J.Y.; Chin, T.S.; Shun, T.T.; Tsau, C.H.; Chang, S.Y. Nanostructured High-Entropy Alloys with Multiple Elements: Novel Alloy Design Concepts and Outcomes. *Adv. Eng. Mater.* **2004**, *6*, 299–303. [CrossRef]
2. Li, W.D.; Liaw, P.K.; Gao, Y. Fracture resistance of high entropy alloys: A review. *Intermetallics* **2018**, *99*, 69–83. [CrossRef]
3. Liu, W.H.; He, J.Y.; Huang, H.L.; Wang, H.; Lu, Z.P.; Liu, C.T. Effects of Nb additions on the microstructure and mechanical property of CoCrFeNi high-entropy alloys. *Intermetallics* **2015**, *60*, 1–8. [CrossRef]
4. He, J.Y.; Liu, W.H.; Wang, H.; Wu, Y.; Liu, X.J.; Nieh, T.G.; Lu, Z.P. Effects of Al addition on structural evolution and tensile properties of the FeCoNiCrMn high-entropy alloy system. *Acta Mater.* **2014**, *62*, 105–113. [CrossRef]
5. Senkov, O.N.; Wilks, G.B.; Scott, J.M.; Miracle, D.B. Mechanical properties of NbMoTaW and VNbMoTaW refractory high entropy alloys. *Intermetallics* **2011**, *19*, 698–706. [CrossRef]
6. Lu, Y.P.; Dong, Y.; Guo, S.; Jiang, L.; Kang, H.J.; Wang, T.M.; Wen, B.; Wang, Z.J.; Jie, J.C.; Ruan, H.H.; et al. A Promising New Class of High-Temperature Alloys: Eutectic High-Entropy Alloys. *Sci. Rep.* **2014**, *4*, 6200. [CrossRef]
7. Gludovatz, B.; Hohenwarter, A.; Catoor, D.; Chang, E.H.; George, E.P.; Ritchie, R.O. A fracture-resistant high-entropy alloy for cryogenic applications. *Science* **2014**, *345*, 1153–1158. [CrossRef]
8. Otto, F.; Dlouh, A.; Somsen, C.H.; Bei, H.; Eggeler, G.; George, E.P. The influences of temperature and microstructure on the tensile properties of a CoCrFeMnNi high-entropy alloy. *Acta Mater.* **2013**, *61*, 5743–5755. [CrossRef]
9. Chuang, M.H.; Tsai, M.H.; Wang, W.R.; Lin, S.J.; Yeh, J.W. Microstructure and wear behavior of $Al_xCo_{1.5}CrFeNi_{1.5}Ti_y$ high-entropy alloys. *Acta Mater.* **2011**, *59*, 6308–6317. [CrossRef]
10. Hemphill, M.A.; Yuan, T.; Wang, G.Y.; Yeh, J.W.; Tsai, C.W.; Chuang, A.; Liaw, P.K. Fatigue behavior of $Al_{0.5}CoCrCuFeNi$ high entropy alloys. *Acta Mater.* **2012**, *60*, 5723–5734. [CrossRef]
11. Lu, Y.P.; Dong, Y.; Jiang, H.; Wang, Z.J.; Cao, Z.Q.; Guo, S.; Wang, T.M.; Li, T.J.; Liaw, P.K. Promising properties and future trend of eutectic high entropy alloys. *Scr. Mater.* **2020**, *187*, 202–209. [CrossRef]
12. Lu, Y.P.; Gao, X.Z.; Jiang, L.; Chen, Z.N.; Wang, T.M.; Jie, J.C.; Kang, H.J.; Zhang, Y.B.; Guo, S.; Ruan, H.H. Directly cast bulk eutectic and near-eutectic high entropy alloys with balanced strength and ductility in a wide temperature range. *Acta Mater.* **2017**, *124*, 143–150. [CrossRef]

13. Wang, M.L.; Lu, Y.P.; Lan, J.G.; Wang, T.M.; Zhang, C.; Cao, Z.Q.; Li, T.J.; Liaw, P.K. Lightweight, ultrastrong and high thermal-stable eutectic high-entropy alloys for elevated-temperature applications. *Acta Mater.* **2023**, *248*, 118806. [CrossRef]
14. Wang, M.L.; Lu, Y.P.; Lan, J.G.; Wang, T.M.; Zhang, C.A.; Cao, Z.Q.; Li, T.J.; Liaw, P.K. A novel bulk eutectic high-entrogy alloy with outstanding as-cast specific yield strengths at elevated temperatures. *Scr. Mater.* **2021**, *204*, 114132. [CrossRef]
15. Ma, S.G.; Jiao, Z.M.; Qiao, J.W.; Yang, H.J.; Zhang, Y.; Wang, Z.H. Strain rate effects on the dynamic mechanical properties of the AlCrCuFeNi2 high-entropy alloy. *Mater. Sci. Eng. A* **2016**, *649*, 35–38. [CrossRef]
16. Dong, Y.; Gao, X.X.; Lu, Y.P.; Wang, T.M.; Li, T.J. A multi-component $AlCrFe_2Ni_2$ alloy with excellent mechanical properties. *Mater. Lett.* **2016**, *169*, 62. [CrossRef]
17. Li, Z.; Zhao, S.; Diao, H.; Liaw, P.K.; Meyers, M.A. High-velocity deformation of $Al_{0.3}CoCrFeNi$ high-entropy alloy: Remarkable resistance to shear failure. *Sci. Rep.* **2017**, *7*, 42742. [CrossRef]
18. Meyers, M.A. (Ed.) *Dynamic Behavior of Materials*; John Wiley and Sons: Hoboken, NJ, USA, 1994.
19. Cho, K.M.; Lee, S.H.; Nutt, S.R.; Duffy, J. Adiabatic shear band formation during dynamic torsional deformation of an HY-100 steel. *Acta Mater.* **1993**, *41*, 923–932. [CrossRef]
20. Lee, H.S.; Sohn, S.S.; Jeon, C.W.; Jo, I.G.; Lee, S.K.; Lee, S.H. Dynamic compressive deformation behavior of SiC-particulate-reinforced A356 Al alloy matrix composites fabricated by liquid pressing process. *Mater. Sci. Eng. A* **2017**, *680*, 368–377. [CrossRef]
21. Dirras, G.; Couque, H.; Lilensten, L.; Heczel, A.; Tingaud, D.; Couzinié, J.-P.; Perriére, L.; Gubicza, J.; Guillot, I. Mechanical behavior and microstructure of Ti20Hf20Zr20Ta20Nb20 high-entropy alloy loaded under quasi-static and dynamic compression conditions. *Mater. Charact.* **2016**, *111*, 106–113. [CrossRef]
22. Wang, L.; Qiao, J.W.; Ma, S.G.; Jiao, Z.M.; Zhang, T.W.; Chen, G.; Zhao, D.; Zhang, Y.; Wang, Z.H. Mechanical response and deformation behavior of $Al_{0.6}CoCrFeNi$ high-entropy alloys upon dynamic loading. *Mater. Sci. Eng.* **2018**, *727*, 208–213. [CrossRef]
23. Tian, L.; Jiao, Z.M.; Yuan, G.Z.; Ma, S.G.; Wang, Z.H.; Yang, H.J.; Zhang, Y.; Qiao, J.W. Effect of Strain Rate on Deformation Behavior of AlCoCrFeNi High-Entropy Alloy by Nanoindentation. *Mater. Eng. Perform.* **2016**, *25*, 2255–2260. [CrossRef]
24. Cao, T.Q.; Zhang, Q.; Wang, L.; Wang, L.; Xiao, Y.; Yao, J.H.; Liu, H.Y.; Ren, Y.; Liang, J.; Xue, Y.F. Dynamic deformation behaviors and mechanisms of CoCrFeNi high-entropy alloys. *Acta Mater.* **2023**, *260*, 119343. [CrossRef]
25. Yuan, K.B.; Yao, X.H.; Yu, Y.Q.; Wang, R.F.; Chai, Z.S.; Zhou, K.X.; Wang, Z.J. Dynamic thermomechanical response and constitutive modeling of eutectic high-entropy alloy. *Int. J. Mech. Sci.* **2023**, *246*, 108148. [CrossRef]
26. Park, J.M.; Moon, J.; Bea, J.W.; Jang, M.J.; Park, J.; Lee, S.; Kim, H.S. Strain rate effects of dynamic compressive deformation on mechanical properties and microstructure of CoCrFeMnNi high-entropy alloy. *Mater. Sci. Eng.* **2018**, *719*, 155–163. [CrossRef]
27. Kumar, N.; Ying, Q.; Nie, X.; Mishra, R.S.; Tang, Z.; Liaw, P.K.; Brenan, R.E.; Doherty, K.J.; Cho, K.C. High strain-rate compressive deformation behavior of the $Al_{0.1}CrFeCoNi$ high entropy alloy. *Mater. Des.* **2015**, *86*, 598–602. [CrossRef]
28. Zhang, Y.; Zuo, T.T.; Tang, Z.; Gao, M.C.; Dahmen, K.A.; Liaw, P.K.; Lu, Z.P. Microstructures and properties of high-entropy alloys. *Prog. Mater. Sci.* **2014**, *61*, 1–93. [CrossRef]
29. Wang, B.F.; Fu, A.; Huang, X.X.; Liu, B.; Liu, Y.; Li, Z.Z.; Zan, X. Mechanical properties and microstructure of the CoCrFeMnNi high entropy alloy under high strain rate compression. *Mater. Eng. Perform.* **2016**, *25*, 2985–2992. [CrossRef]
30. Dong, Y.; Zhou, K.Y.; Lu, Y.P.; Gao, X.X.; Wang, T.M.; Li, T.J. Effect of vanadium addition on the microstructure and properties of AlCoCrFeNi high entropy alloy. *Mater. Des.* **2014**, *57*, 67–72. [CrossRef]
31. Singh, S.; Wanderka, N.; Murty, B.S.; Glatzel, U.; Banhart, J. Decomposition in multi-component AlCoCrCuFeNi high-entropy alloy. *Acta Mater.* **2011**, *59*, 182–190. [CrossRef]
32. Santodonato, L.J.; Zhang, Y.; Feygenson, M.; Parish, C.M.; Gao, M.C.; Weber, R.J.K.; Neuefeind, J.C.; Tang, Z.; Liaw, P.K. Deviation from high-entropy configurations in the atomic distributions of a multi-principal-element alloy. *Nat. Commun.* **2015**, *6*, 5964. [CrossRef]
33. Linden, Y.; Pinkas, M.; Munitz, A.; Meshi, L. Long-period antiphase domains and short-range order in a B2 matrix of the AlCoCrFeNi high-entropy alloy. *Scr. Mater.* **2017**, *139*, 49–52. [CrossRef]
34. Moon, J.; Hong, S.I.; Bae, J.W.; Jang, M.J.; Yim, D.; Kim, H.S. On the strain rate-dependent deformation mechanism of CoCrFeMnNi high-entropy alloy at liquid nitrogen temperature. *Mater. Res. Lett.* **2017**, *5*, 472–477. [CrossRef]
35. Tang, Y.; Wang, R.X.; Xiao, B.; Zhang, Z.R.; Li, S.; Qiao, J.W.; Bai, S.X.; Zhang, Y.; Liaw, P.K. A review on dynamic-mechanical behaviors of high entropy alloys. *Prog. Mater. Sci.* **2023**, *135*, 101090. [CrossRef]
36. Hong, S.I.; Moon, J.; Hong, S.K.; Kim, H.S. Thermally activated deformation and the rate controlling mechanism in CoCrFeMnNi high entropy alloy. *Mater. Sci. Eng. A* **2017**, *682*, 569–576. [CrossRef]
37. Hines, J.A.; Vecchio, K.S. Recrystallization kinetics within adiabatic shear bands. *Acta Mater.* **1997**, *45*, 635–649. [CrossRef]
38. Chen, W.; Song, B. *Split Hopkinson (Kolsky) Bar*; Springer: New York, NY, USA, 2011.
39. Conque, H. The use of the direct impact Hopkinson pressure bar technique to describe thermally activated and viscous regimes of metallic materials. *Philos. Trans. A Math. Phys. Eng. Sci* **2014**, *372*, 20130218.

40. Meyers, M.A.; Nesterenko, V.F.; Lasalvis, J.C.; Xue, Q. Shear localization in dynamic deformation of materials: Microstructural evolution and self-organization. *Mater. Sci. Eng. A* **2001**, *317*, 204–225. [CrossRef]
41. Ahn, D.H.; Lee, D.J.; Kang, M.J.; Park, L.J.; Lee, S.H.; Kim, H.S. Bi-modal structure of copper via room-temperature partial recrystallization after cryogenic dynamic compression. *Metall. Mater. Trans. A* **2016**, *47*, 1600–1606. [CrossRef]
42. Bai, J.; Kou, H.C.; Wang, J.; Li, J.S.; Hu, R. Strain rate response of a Ti-based metallic glass composite at cryogenic temperature. *Mater. Lett.* **2014**, *117*, 228–230. [CrossRef]
43. Jiang, K.; Ren, T.F.; Shan, G.B.; Ye, T.; Chen, L.Y.; Wang, C.X.; Zhao, F.; Li, J.G.; Suo, T. Dynamic mechanical responses of the $Al_{0.1}CoCrFeNi$ high entropy alloy at cryogenic temperature. *Mater. Sci. Eng. A* **2020**, *797*, 140125. [CrossRef]

Disclaimer/Publisher's Note: The statements, opinions and data contained in all publications are solely those of the individual author(s) and contributor(s) and not of MDPI and/or the editor(s). MDPI and/or the editor(s) disclaim responsibility for any injury to people or property resulting from any ideas, methods, instructions or products referred to in the content.

materials

MDPI

Article

Synergistic Effect of Cu Addition and Pre-Straining on the Natural Aging and Artificial Age-Hardening Behavior of AA6111 Alloy

Shougang Duan [1], Yizhe Lu [1], Aiwen Li [1], Mingkan Tang [1], Weilun Chen [1], Chengyi Huang [1,2,3], Jun Du [1,*], Yanping Xu [2,3] and Yan Yan [2,3]

1 School of Materials Science and Engineering, South China University of Technology, Guangzhou 510640, China; sgduan0410@163.com (S.D.); l995709739@163.com (Y.L.); 202411085587@mail.scut.edu.cn (A.L.); 19125629185@163.com (M.T.); cwl13924509313@163.com (W.C.); huangcehngyi@algig.cn (C.H.)

2 Algig Aluminum Inc., Nanning 530031, China; xuyanping@algig.cn (Y.X.); yanyan@algig.cn (Y.Y.)

3 Guangxi Key Laboratory of Materials and Processes of Aluminum Alloys, Nanning 530031, China

* Correspondence: tandujun@sina.com

Abstract: This study systematically investigates the synergistic effects of Cu addition (0–0.7 wt.%) and 2% pre-straining on the artificial aging, natural aging (NA), and bake-hardening response (BHR) of AA6111 alloy. The results reveal that Cu significantly enhances age-hardening capacity and accelerates artificial aging kinetics. The 0.7Cu alloy achieved a 14% higher peak hardness (106.9 HV) than the Cu-free alloy (93.8 HV) while reducing peak aging time by 50% (from 10 h to 5 h). Pre-straining further promoted hardening rates, shortening peak aging times to 2 h for the 0.7Cu alloy. Natural aging (NA) severely suppressed BHR in non-pre-strained alloys, reducing paint baking (PB) increments by 75–77.5% after 14 days. However, the introduction of pre-straining not only reduces the negative effects of NA but also improves the BHR. TEM analysis demonstrated that Cu addition accelerated the precipitation of fine GP zones and β'' phases while pre-straining introduced dislocations that acted as heterogeneous nucleation sites for Q' phases, refining precipitates and suppressing NA cluster formation. The synergistic combination of 0.7Cu and pre-straining achieved a superior BHR yield strength increment of 68.1 MPa with retained ductility, highlighting its potential for automotive applications requiring balanced formability and post-forming strength.

Keywords: Al-Mg-Si-(Cu) alloys; pre-straining; Cu addition; age-hardening response; precipitation

1. Introduction

In recent years, with the acceleration of automotive lightweighting, 6000-series (Al-Mg-Si-Cu) aluminum (Al) alloys have become the preferred material for body-panel components due to their excellent strength–formability balance and good bake-hardening response (BHR) [1–4]. Among them, the AA6111 alloy sheet has attracted widespread attention from both academia and industry because of its significant strength-enhancement characteristics during the paint baking (PB) process after stamping forming [1,5]. The performance optimization of this alloy is highly dependent on the structural evolution of nanoscale precipitates during artificial aging, and copper (Cu), as a key alloying element, has a decisive impact on the alloy's aging kinetics and final mechanical properties by changing the precipitation sequence and regulating the type and distribution of precipitates [6–9]. However, a persistent challenge lies in the adverse effects of NA during room-temperature

https://doi.org/10.3390/ma18071635

storage, which leads to premature cluster formation, thereby consuming solute atoms and diminishing the alloy's capacity for subsequent artificial aging [10,11]. This phenomenon severely compromises BHR, limiting the alloy's ability to achieve target strength after PB cycles.

In actual industrial production, two methods are commonly used to suppress the negative effects of natural aging (NA): micro-alloying and immediate pre-treatment after solution quenching [12–15]. Cu is one of the most important micro-alloying elements in Al-Mg-Si alloys [4,6,7,16–20]. The addition of Cu can affect the aging behavior of Al-Mg-Si-based alloys through two mechanisms: firstly, Cu atoms enter the β'' phase to form a more stable L/Q′ phase [21], delaying the over-aging process; secondly, it promotes the precipitation density in the early stage of low-temperature aging by reducing the nucleation barrier of GP zones [6,8]. However, Cu alone cannot effectively reduce the negative effects of NA in alloy sheets. Zandbergen et al. [18] found that alloys with higher Cu content had higher natural aging hardening rates and believed that the addition of Cu would reduce the strain energy of natural aging clusters, thereby releasing more vacancies and promoting the growth of NA clusters. Recent studies have indicated that thermo-mechanical processing, particularly pre-straining (PS), can suppress the NA effect by introducing dislocations that capture quenched vacancies and altering the precipitation path [10,12]. Furthermore, pre-straining can alter the structural types of precipitates. For instance, Weng et al. [22] found that precipitates formed along dislocations can exhibit three characteristics: non-periodic atomic arrangement within the precipitates; segregation of Cu at the precipitate/α (Al) interface; and different orientations presented in a single precipitate.

Studies have shown that the interaction between Cu and dislocations is beneficial to the aging properties of 6xxx-series Al alloys. Ding et al. [23] found that the addition of 0.5 wt.% Cu enhances the effect of pre-straining on improving BHR by promoting the preferential precipitation of the Q′ phase. In contrast, Saito et al. [24] reported that pre-straining did not alter the hardness or precipitate types of a low-Cu (0.1 wt.%) alloy aged at 175 °C for 300 min. These findings indicate that the influence of dislocations on aging behavior is closely related to alloy composition, particularly Cu content. Although the effects of pre-straining have been extensively studied, most research has focused on low-Cu 6xxx alloys (e.g., 6016, 6022). For the high-Cu 6111 alloy, the synergistic interaction mechanisms between Cu and pre-straining remain have not been deeply studied. Elucidating the cooperative effects of Cu and pre-straining on the aging performance of 6xxx series alloys is critical for optimizing their mechanical properties. Therefore, this study systematically investigates the coupled effects of Cu content (0–0.7 wt.%) and 2% pre-straining on the artificial aging behavior, natural aging, and tensile performance of AA6111 alloys. Through hardness testing, tensile characterization, and microstructural analysis (TEM/HRTEM), we elucidate how Cu accelerates precipitation kinetics while pre-straining modulates dislocation-mediated phase nucleation. We provide a mechanistic framework for optimizing 6xxx alloys via composition-process synergy, advancing their applicability in high-performance automotive manufacturing.

2. Experimental

Table 1 presents the chemical compositions of three experimental Al-Mg-Si-Cu alloys with varying Cu contents (designated as 0Cu, 0.2Cu, and 0.7Cu). The alloys were prepared using industrial pure Al (99.7 wt.%) and master alloys containing Al-10 wt.% Mg, Al-20 wt.% Si, Al-50 wt.% Cu, Al-20 wt.% Fe, and Al-10 wt.% Mn. Precisely weighed raw materials were melted in a graphite crucible using a resistance furnace. To minimize oxidation, the furnace chamber was purged with high-purity argon gas (99.8%) prior to heating, and a continuous argon flow (0.5 L/min) was maintained throughout the melting process. Melting was conducted at 740 ± 10 °C for 40 min with intermittent mechanical stirring to ensure compositional homogeneity. The melt was then poured into a preheated (250 °C) cast steel mold with a rectangular cavity (dimensions: 100 mm × 150 mm × 20 mm) using a tilt-pouring technique. The entire process was completed within an argon-protected environment to prevent gas entrapment and oxide formation. Chemical composition verification was performed through optical emission spectroscopy (HITACHI OE750, HITACHI, Tokyo, Japan) prior to subsequent processing. The as-cast ingots underwent homogenization treatment at 560 °C for 8 h in a forced-air circulation furnace, followed by furnace cooling to an ambient temperature. Surface preparation included mechanical milling to remove 1 mm from each face, eliminating surface oxidation and casting defects. Thermomechanical processing consisted of two stages: hot rolling from 18 mm to 10 mm thickness at 460 °C (reduction rate: 0.8 mm/pass), followed by cold rolling to final 4 mm sheets (total reduction: 60%). Solution treatment was conducted at 520 °C for 30 min in a salt bath, with immediate water quenching to room temperature. Post-quench processing involved two experimental conditions: (1) direct NA for 14 days at room temperature and (2) the application of 2% tensile pre-straining (PS_2) prior to NA. All specimens subsequently underwent artificial aging at 180 °C for durations ranging from 0 to 24 h. Additionally, a simulated paint bake cycle (180 °C/30 min) was applied to selected samples during natural aging. The complete experimental sequence is schematically illustrated in Figure 1a.

Table 1. Chemical composition of the three investigated alloys (wt.%).

	Mg	Si	Cu	Mn	Fe	Al
0Cu	0.76	0.85	0.01	0.28	0.16	Bal.
0.2Cu	0.72	0.86	0.21	0.28	0.15	Bal.
0.7Cu	0.75	0.86	0.68	0.29	0.16	Bal.

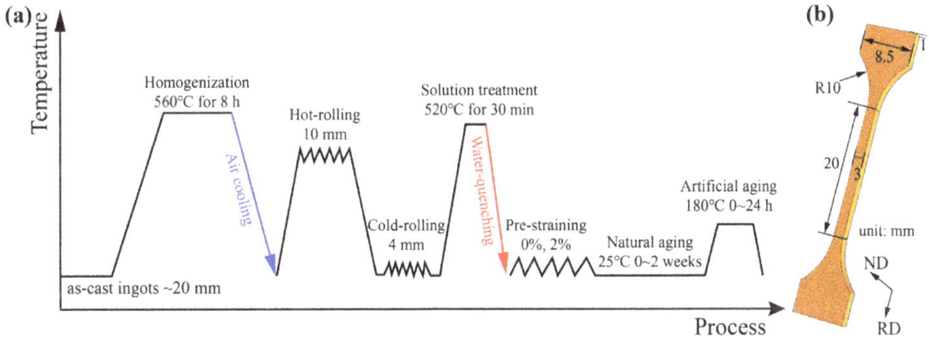

Figure 1. Schematic diagram of (**a**) the process of heat treatment and (**b**) the tensile specimen size.

Vickers hardness measurements were conducted using an HVS-1000Z microhardness tester (Caikon, Shanghai, China) under a 500 gf load with 10 s dwell time. Seven indentations per condition were systematically recorded, with the average value and standard deviation calculated to ensure data reliability. Uniaxial tensile testing was performed on room temperature specimens using a Shimadzu AG-X 100 kN universal testing machine (Kyoto, Japan) with a constant crosshead speed corresponding to an initial strain rate of 1.0×10^{-3} s^{-1}. Three replicate tests per condition were conducted following ASTM E8 standards [25], with the specimen geometry details illustrated in Figure 1b. Microstructural evolution was investigated using a Thermo-Fisher (Waltham, MA, USA) Talos F200X field emission transmission electron microscope (TEM). All TEM and high-resolution TEM (HRTEM) images in this study were acquired along the <100>$_{Al}$ zone axis. TEM samples were prepared by a combination of twin-jet electron polishing with ion thinning. Firstly, the 3 mm diameter disks, which had been mechanically ground to 80 ± 5 µm, were subjected to twin-jet electrolytic thinning using a cryogenic electrolyte (30 vol.% HNO_3 in methanol, −25 ± 1 °C) under 25.0 ± 2 V potential. Final thinning was achieved using a Gatan 691 Precision Ion Polishing System with low-angle (4 ± 0.5°) 3 keV Ar^+ bombardment.

3. Results

3.1. Artificial Age-Hardening Behavior

Figure 2 illustrates the artificial aging hardness evolution (180 °C, 0–24 h) of three experimental alloys in both pre-strained and non-pre-strained conditions after two weeks of room-temperature storage. Among the three non-pre-strained alloys, the 0.7Cu alloy exhibited the highest hardness throughout the artificial aging process, while the Cu-free alloy consistently had the lowest hardness (Figure 2a). Additionally, the 0.7Cu alloy consistently exhibited the fastest artificial aging response rate before reaching peak aging. The peak hardness values of the 0.2Cu and 0.7Cu alloys were measured to be 103.1 HV and 106.9 HV, respectively. These values represent increases of 9.9% and 14% compared to the peak hardness of the 0Cu alloy, which registered at 93.8 HV. Regarding the time required to reach peak aging, the 0Cu alloy took 11 h, while the 0.7Cu alloy required only 5 h, signifying a 55% reduction. The peak aging time for the 0.2Cu alloy was also reduced to 7 h. These results demonstrate that Cu addition not only enhances age-hardening capacity but also significantly reduces the time to peak aging.

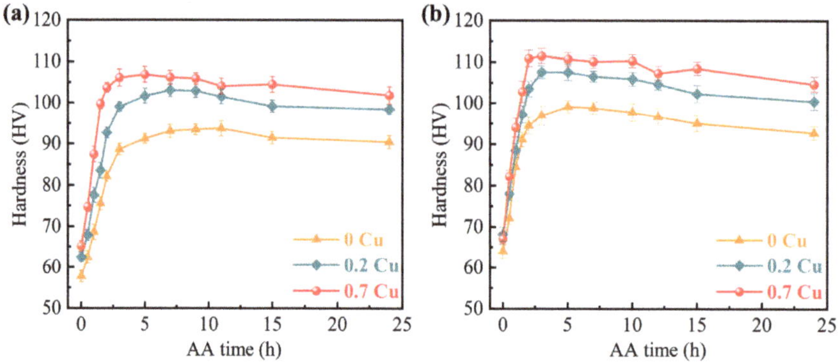

Figure 2. Vickers hardness evolution of the three investigated alloys during 24 h artificial aging at 180 °C, with and without pre-straining. (**a**) Without pre-straining; (**b**) with pre-straining to 2%.

After a pre-straining of 2%, the three investigated alloys exhibited similar hardening rates in the early artificial aging period (Figure 2b). In comparison to the without pre-straining alloys, the 0.7Cu alloy maintained the highest hardness throughout the artificial aging process, while the 0Cu alloy exhibited the lowest hardness. The peak hardnesses of the 0Cu, 0.2Cu, and 0.7Cu alloys were 99.2 HV, 107.6 HV, and 111.7 HV, respectively, which were 5.4 HV, 4.5 HV, and 4.8 HV higher than those of the non-pre-strained alloys. More importantly, pre-straining significantly reduces the time required to reach peak aging. The peak aging times were reduced to 7 h, 5 h, and 2 h for the 0Cu, 0.2Cu, and 0.7Cu alloys, respectively. This result implies that the introduction of pre-straining effectively promotes the age-hardening behavior of the alloys. In particular, this enhancement is more pronounced in the 0Cu alloy, suggesting potential differences in its response to pre-straining as compared to the 0.2Cu and 0.7Cu alloys.

3.2. Natural Aging and Bake Hardening Response

Figure 3 demonstrates the NA behavior and PB hardening increments (ΔHV) of AA6111 alloys with varying Cu content (0Cu, 0.2Cu, 0.7Cu) under pre-strained and non-pre-strained conditions during natural aging. The base hardness of the alloys exhibited a progressive enhancement with increasing Cu content. For non-pre-strained samples (Figure 3a–c), a monotonic hardness increase was observed across all alloys as NA duration extended from 0 to 14 days, while the PB increment demonstrated an inverse relationship, decreasing by 75.6% (0Cu), 75.6% (0.2Cu), and 77.5% (0.7Cu), respectively. The 0.7Cu alloy consistently demonstrated superior BHR compared to the Cu-free counterpart throughout the 0–14 days of NA, underscoring the critical role of Cu in retaining bake-hardening potential. Freshly solution-treated samples (without NA) exhibited maximum PB increments, with 0Cu, 0.2Cu, and 0.7Cu alloys achieving 12.3 HV, 18 HV, and 22.7 HV, respectively. After 14-day NA, the hardness of the 0Cu alloy increased from 42.8 HV to 58.9 HV, while the PB increment was only 3 HV (Figure 3a), which was significantly lower than that of the direct PB sample (ΔHV = 12.3 HV). Similarly, the hardness of the 0.2Cu and 0.7Cu alloys increased from 57.3 HV to 68.5 HV and 54.5 HV to 67.1 HV, respectively, while the PB increments were only 4.4 HV and 5.1 HV. Accordingly, although the addition of Cu could enhance the BHR of AA6111 alloy, it is difficult to effectively suppress the negative effects of NA.

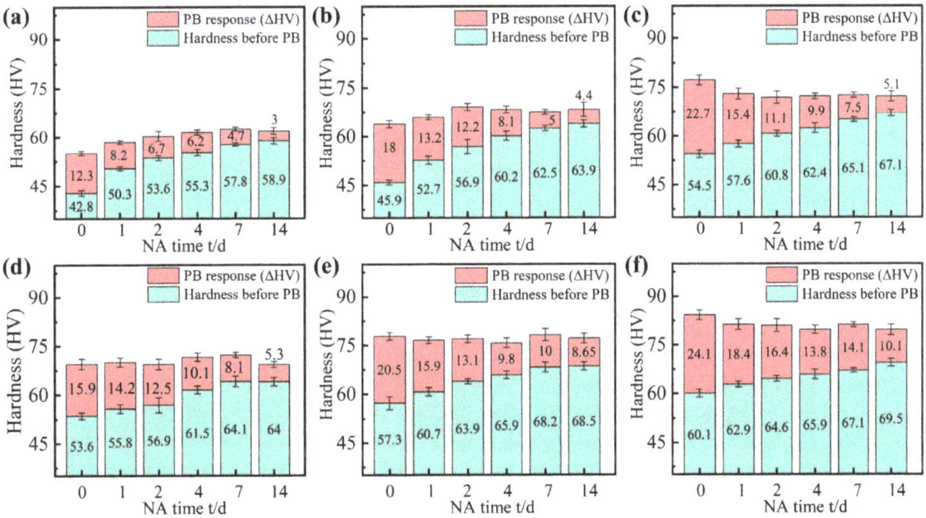

Figure 3. Vickers hardness evolution during two weeks of RT storage and with PB treatment. (**a**–**c**) Non-pre-strained alloys and (**d**–**f**) alloys with 2% pre-straining; (**a**,**d**) the 0Cu alloy, (**b**,**e**) the 0.2Cu alloy, and (**c**,**f**) the 0.7Cu alloy.

The strain hardening resulting from the introduction of 2% pre-straining increased the hardness of all three alloys (Figure 3d–f). For the three pre-strained alloys, the incremental hardness due to ($ΔHV_{NA}$) was reduced after two weeks in natural parking. The $ΔHV_{NA}$ for the 0Cu, 0.2Cu, and 0.7Cu alloys was reduced from 16.1 HV to 10.4 HV, 18 HV to 11.2 HV, and 12.6 HV to 9.4 HV, respectively. Furthermore, the introduction of pre-strain effectively enhances the BHR of the AA6111 alloy. These results indicate increasing the Cu content in combination with pre-straining not only suppresses the negative effects of NA but also enhances the BHR.

3.3. Tensile Response

Figure 4 presents the engineering tensile stress–strain curves of the three experimental alloys with and without 2% pre-straining, measured before and after PB treatment (180 °C/30 min) following 14-day NA. The quasi-static tensile tests systematically revealed the synergistic effects of pre-straining and Cu alloying on the plastic deformation mechanisms of the AA6111 series Al alloys. The three non-pre-strained alloys exhibited relatively low initial yield strengths (Figure 4a), with limited improvements observed after PB treatment (Figure 4b). Table 2 lists detailed values for yield strength, ultimate tensile strength (UTS), and elongation (El). Quantitative analysis demonstrated that the yield strength increments for 0Cu, 0.2Cu, and 0.7Cu alloys were 12.8 MPa, 14.2 MPa, and 17.7 MPa, respectively, consistent with the hardness results shown in Figure 3. With an increasing Cu content, the alloy strength was enhanced while retaining excellent elongation. After applying 2% pre-straining (Figure 4c,d), all alloys exhibited increased strength with slight reductions in elongation. Notably, pre-straining enhanced the bake-hardening yield strength increment more substantially. The PB-induced yield strength increments for 0Cu, 0.2Cu, and 0.7Cu alloys reached 29.8 MPa, 41.3 MPa, and 68.1 MPa, respectively. In particular, the 0.7Cu alloy not only demonstrated a markedly improved bake-hardening yield strength increment but also maintained satisfactory elongation after pre-straining.

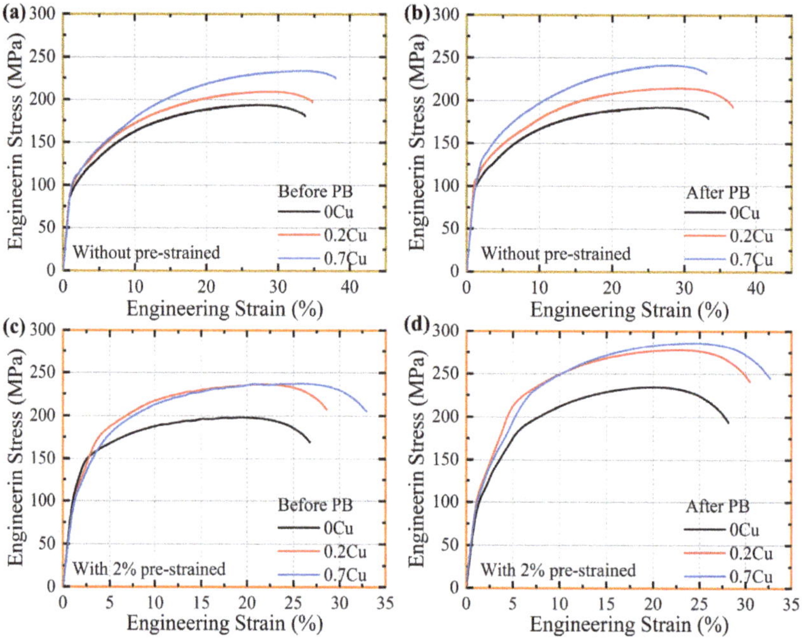

Figure 4. Stress–strain curves (**a**,**c**) before and (**b**,**d**) after PB treatment for three alloys. (**a**,**b**) Non-pre-straining and (**c**,**d**) 2% pre-straining.

Table 2. Tensile properties of the three investigated alloys with and without pre-straining.

Alloys	PS_x	YS (MPa)		UTS (MPa)		El (%)		BHR (Δ_{YS}, MPa)
		Before PB	After PB	Before PB	After PB	Before PB	After PB	
0Cu	PS_0	91.4	104.2	193.6	191.9	33.6	33.3	12.8
	PS_2	151.4	181.2	197.8	234.4	26.8	28.2	29.8
0.2Cu	PS_0	99.6	113.8	209.3	214.6	34.7	36.8	14.2
	PS_2	176.5	217.8	236.5	278.3	28.6	30.5	41.3
0.7Cu	PS_0	102.9	120.6	233.8	241.3	37.9	32.9	17.7
	PS_2	156.8	224.9	237.3	286.1	32.9	32.6	68.1

3.4. Microstructural Characterization

Figure 5 shows the bright-field (BF) TEM images of the 0Cu alloy and 0.7Cu alloy without pre-straining, both without NA and after 14 days of NA, followed by aging at 180 °C for 30 min. All TEM images were taken along the $<100>_{Al}$ zone axis. For the 0Cu alloy, in the absence of NA, there are a few needle-like precipitates and uniformly distributed fine dot-like precipitates (Figure 5a). Following 14 days of NA, both the size reduction in needle-shaped precipitates and the marked decrease in density of dot-like precipitates indicate intensified detrimental effects of NA, leading to a diminished BHR response in the alloy (Figure 5a). High-resolution TEM (HRTEM) imaging and the corresponding Fast Fourier Transform (FFT) analysis in Figure 6 confirm that the smaller dot-like precipitates correspond to GP zones, while larger ones are identified as β'' phase. In the case of the 0.7Cu alloy (Figure 5c,d), a significantly higher density of dot-like precipitates is uniformly distributed within the Al matrix before NA compared to the Cu-free counterpart, demonstrating that Cu addition accelerates the precipitation of aging–strengthening phases. This aligns with the hardness evolution trends shown in Figure 3. Notably, after 14 days of NA, the post-PB precipitate density in the 0.7Cu alloy also decreases, further confirming that Cu incorporation alone cannot counteract the negative effects induced by NA.

To further investigate the effect of pre-straining on the precipitation behavior of the alloy, TEM and HRTEM characterizations were conducted on the 0.7Cu alloy subjected to 2% pre-straining, as shown in Figures 7 and 8. As shown in Figure 7a, a high density of dislocations introduced by pre-straining was observed in the low-magnification BF-TEM image, with numerous precipitates distributed throughout the α-Al matrix. The high-magnification BF-TEM image in Figure 7b revealed that dark dot-shaped precipitates and light needle-shaped precipitates are predominantly distributed near dislocations and within the α-Al matrix. The needle-shaped precipitates correspond to the primary strengthening phase β'', while most dot-shaped precipitates consist of GP zones and the β'' phase, along with a small amount of Q' phase associated with high Cu content. After 14 days of NA followed by PB treatment, as shown in Figure 7c,d, the 0.7Cu alloy matrix still retained numerous dot-shaped and needle-shaped phases. This indicates that pre-straining significantly inhibited the negative effects of NA. Compared with non-pre-strained samples (Figure 5), the needle-shaped precipitates exhibited greater length and higher density, confirming that pre-straining not only effectively suppresses NA's detrimental effects but also enhances the alloy's BHR capability. Following pre-straining, the precipitates growing along dislocation lines displayed continuous, elongated characteristics. Furthermore, these precipitates exhibited disordered crystal structures, as supported by the HRTEM image and corresponding FFT pattern shown in Figure 8.

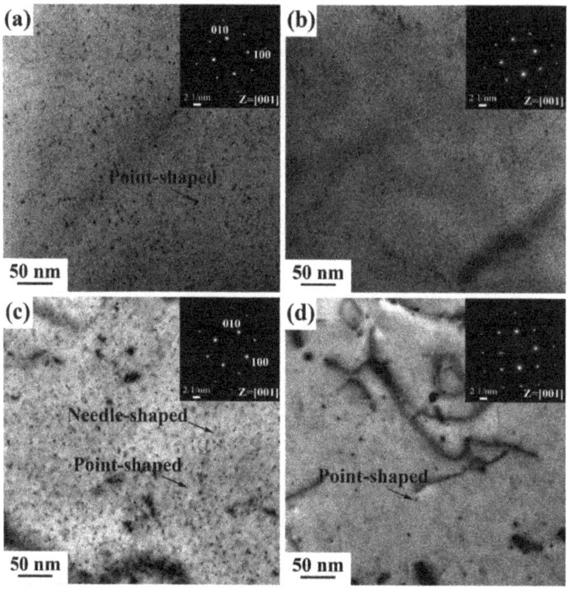

Figure 5. Bright-field TEM images of 0Cu (**a**,**b**) and 0.7Cu (**c**,**d**) alloys after PB without pre-straining. (**a**,**c**) Without NA; (**b**,**d**) underwent 14 days of NA.

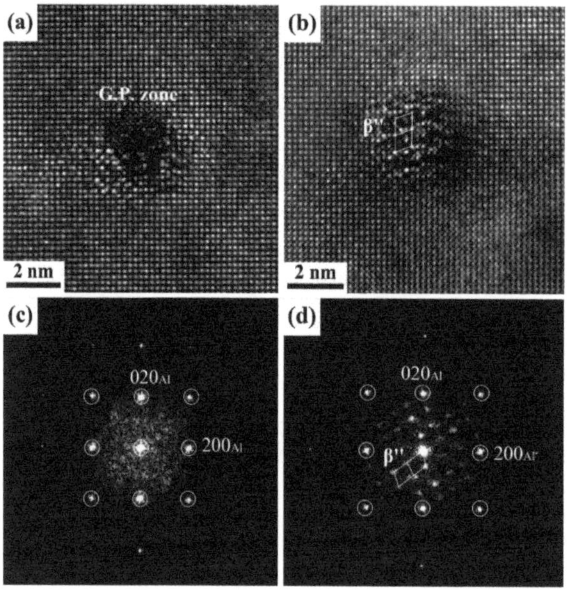

Figure 6. HRTEM images and corresponding FFT patterns of the dot-shaped precipitates in the non-pre-strained alloys. (**a**,**c**) GP zone from 0Cu alloy (0 d); (**b**,**d**) β″ phase from 0.7Cu alloy (14 d).

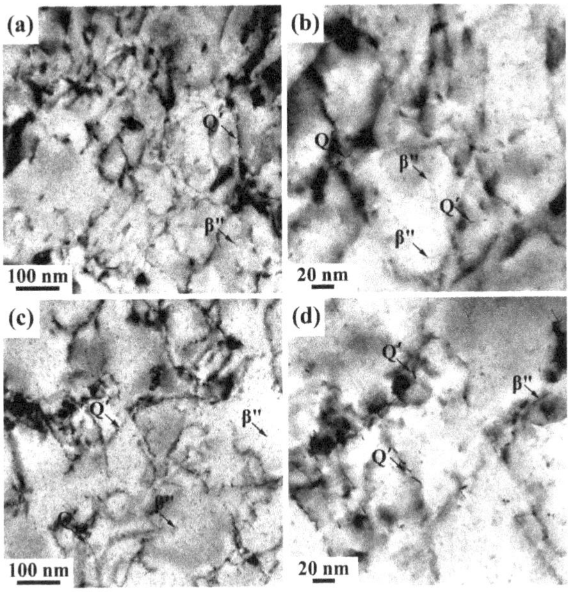

Figure 7. Bright-field TEM images of 0.7Cu alloy with 2% pre-straining subjected to PB treatment after different times of NA. (**a**,**b**) Without previous NA; (**c**,**d**) NA 14 days.

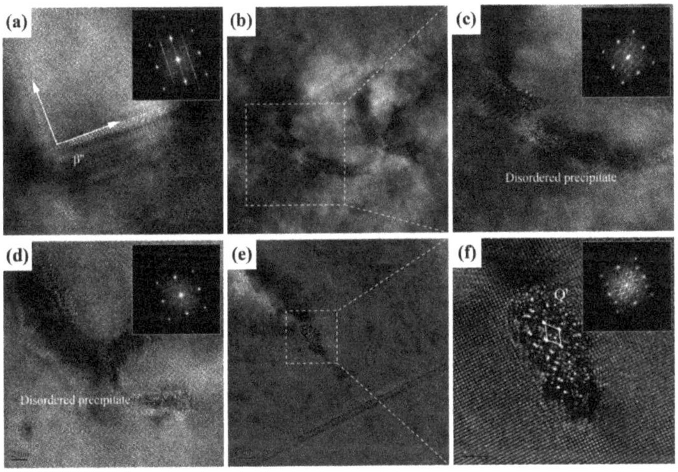

Figure 8. HRTEM images of typical precipitates in 0.7Cu alloy with 2% pre-straining. (**a**) Mutually perpendicular β″ phase; (**b**–**d**) precipitates along dislocation lines; (**e**,**f**) Q′ phase.

4. Discussion

For 6xxx series Al alloy sheets used in automobiles, BHR and resistance to NA effects are two critical performance indicators [26–28]. Superior NA resistance effectively reduces the difficulty of stamping processes for structural components, whereas an enhanced BHR improves the strength of final structural parts [29]. In this study, we investigated the effect of Cu addition on the artificial aging response of the AA6111 alloy. The results demonstrated that increasing the Cu content significantly accelerated the hardening rate and shortened the time to peak aging (Figure 2a). During natural aging, Cu atoms are incorporated into the Mg-Si co-clusters and form Mg-Si-Cu co-clusters [30], thereby increasing the

(Mg + Cu)/Si ratio and promoting the formation of Clusters (2) [31,32]. Additionally, prior studies suggest that Cu addition lowers the activation energy of precipitate nucleation [33]. Consequently, during subsequent artificial aging, Cu addition effectively promotes the direct transformation of more Clusters (2) into GP zones and β'' and Q' phases, resulting in accelerated aging kinetics. With an increasing Cu content, the hardness of solution-treated alloys increases from 57.8 HV for 0Cu to 65.2 HV for 0.7Cu, attributed to solid solution strengthening caused by Cu dissolution in the α-Al matrix. The peak-aged hardness increases from 93.8 HV for 0Cu to 106.9 HV for 0.7Cu, owing to the combined effects of solid solution strengthening and precipitation hardening during subsequent artificial aging. In non-pre-strained alloys, the 0.7Cu alloy exhibited smaller-sized and higher-density precipitates after paint baking (PB) compared to the 0Cu alloy (Figure 5), indicating that Cu-induced microstructure refinement contributes to enhanced precipitation hardening.

While Cu addition alone fails to effectively suppress the adverse effects of natural aging (NA) in alloy sheets, this study further explores the combined influence of pre-straining and high Cu content on NA resistance and artificial aging behavior in the AA6111 alloy. Following 2% pre-straining, the hardness increase during NA in the T4P condition was markedly suppressed (Figure 3). Dislocations introduced by pre-straining trapped quenched vacancies, thereby inhibiting the formation of NA clusters [22,34], with the 0.7Cu alloy exhibiting the most pronounced suppression. During the subsequent PB process, strain fields near dislocations provided heterogeneous nucleation sites for GP zones and enhanced atomic diffusion pathways [35], accelerating the precipitation of strengthening phases. Consistent with prior reports, dislocations in high-Cu alloys promoted the preferential nucleation of the Q' phase during artificial aging (Figure 7), attributed to the propensity of the Q' phase to form at defect sites such as dislocations and grain boundaries [6,31]. Notably, precipitates near dislocations exhibited elongated, disordered morphologies, whereas those in dislocation-free regions adopted needle-like (β'') or lath-shaped (Q') structures [26,36]. This contrast arises from the high interfacial energy and enhanced diffusion near dislocations, which favor directional growth along dislocation lines, while local lattice distortions induce structural disorder [22]. In conclusion, the synergistic effect of Cu and pre-straining enhances the BHR of AA6111 alloy and suppresses the negative effects of NA.

5. Conclusions

The findings of this paper provide important insights into the microstructure-property relationships of Cu-modified 6xxx alloys, emphasizing the importance of coupling composition and thermo-mechanical processing in the design of advanced aluminum sheets. The main conclusions are as follows:

(1) Increasing Cu content (0–0.7 wt.%) in AA6111 alloys enhances age-hardening capacity and accelerates artificial aging kinetics. The 0.7Cu alloy exhibited a 14% higher peak hardness (106.9 HV) and 50% shorter peak aging time (5 h vs. 10 h for the Cu-free alloy), attributed to the accelerated precipitation of GP zones, β'', and Q' phases.

(2) Introducing 2% pre-straining significantly shortened peak aging times (e.g., 2 h for 0.7Cu alloy) and suppressed natural aging (NA) effects by aggregating quenched vacancies and inhibiting NA cluster formation. The pre-strained 0.7Cu alloy achieved a 68.1 MPa yield strength increment after paint baking, outperforming non-pre-strained alloys.

(3) TEM analysis revealed that Cu promotes fine, dense precipitates (GP zones/β''), while pre-straining introduces dislocations that enhance Q' phase nucleation. Dislocation strain fields facilitated elongated, disordered precipitates along defect sites, improving the BHR and NA resistance.

(4) The synergy of 0.7Cu and pre-straining optimizes AA6111 alloys for automotive applications, balancing a high bake-hardening response (111.7 HV peak hardness) with suppressed NA instability. This combination addresses the trade-off between formability (pre-strain tolerance) and post-forming strength, offering a viable strategy for lightweight component manufacturing.

Author Contributions: Conceptualization, S.D., A.L., W.C. and C.H.; Methodology, S.D., Y.L., A.L., M.T. and W.C.; Validation, Y.X.; Formal analysis, S.D., M.T. and W.C.; Investigation, S.D., Y.L., A.L., C.H., Y.X. and Y.Y.; Resources, C.H. and Y.Y.; Writing—original draft, S.D.; Writing—review & editing, J.D.; Supervision, J.D.; Project administration, J.D. and Y.Y.; Funding acquisition, J.D. All authors have read and agreed to the published version of the manuscript.

Funding: This research were funded by Nanning major science and technology project grant number 20221017 and funded by Science and Technology Major Program of Guangxi grant number AA23062026.

Institutional Review Board Statement: Not applicable.

Informed Consent Statement: Not applicable.

Data Availability Statement: The original contributions presented in this study are included in the article. Further inquiries can be directed to the corresponding author.

Conflicts of Interest: Authors Chengyi Huang, Yanping Xu and Yan Yan were employed by the company Algig Aluminum Inc. The remaining authors declare that the research was conducted in the absence of any commercial or financial relationships that could be construed as a potential conflict of interest.

References

1. Gong, W.Y.; Xie, M.J.; Zhang, J.S. Giant bake hardening response of multi-scale precipitation strengthened Al-Mg-Si-Cu-Zn alloy via pre-aging treatments. *Mater. Charact.* **2021**, *181*, 111464.
2. Li, S.S.; Yue, X.; Li, Q.Y.; Peng, H.L.; Dong, B.X.; Liu, T.S.; Hong, Y.Y.; Jun, F.; Shi, L.S.; Feng, Q.; et al. Development and applications of aluminum alloys for aerospace industry. *J. Mater. Res. Technol.* **2023**, *27*, 944–983.
3. Li, T.; Song, J.; Zhang, A.; You, G.Q.; Yang, Y.; Jiang, B.; Qin, X.Y.; Xu, C.; Pan, F.S. Progress and prospects in Mg-alloy super-sized high pressure die casting for automotive structural components. *J. Magnes. Alloy* **2023**, *11*, 4166–4180.
4. Ponka, B. Structure and Mechanical Properties of AlMgSi(Cu) Extrudates Straightened with Dynamic Deformation. *Materials* **2024**, *17*, 3983. [CrossRef]
5. Macwan, A.; Kumar, A.; Chen, D.L. Ultrasonic spot welded 6111-T4 aluminum alloy to galvanized high-strength low-alloy steel: Microstructure and mechanical properties. *Mater. Des.* **2017**, *113*, 284–296.
6. Ding, L.P.; Jia, Z.; Nie, J.; Weng, Y.; Cao, L.; Chen, H.; Wu, X.Z.; Liu, Q. The structural and compositional evolution of precipitates in Al-Mg-Si-Cu alloy. *Acta Mater.* **2018**, *145*, 437–450.
7. Torsæter, M.; Lefebvre, W.; Marioara, C.D.; Andersen, S.J.; Walmsley, J.C.; Holmestad, R. Study of intergrown L and Q′ precipitates in Al–Mg–Si–Cu alloys. *Scr. Mater.* **2011**, *64*, 817–820.
8. Weng, Y.Y.; Jia, Z.H.; Ding, L.P.; Du, K.; Duan, H.C.; Liu, Q.; Wu, X.H. Special segregation of Cu on the habit plane of lath-like β′ and QP2 precipitates in Al-Mg-Si-Cu alloys. *Scr. Mater.* **2018**, *151*, 33–37.
9. Weng, Y.Y.; Li, R.M.; Zheng, L.X.; Zhao, R.X.; Zhou, Y.X.; Chen, J.H.; Ding, L.P. Effects of pre-deformation on microstructure evolution and hemming performance for Al-Mg-Si-Cu alloys. *J. Mater. Res. Technol.* **2025**, *35*, 2840–2851.
10. Cui, M.; Jo, Y.H.; Kim, Y.; Kim., H.; Lee., J. The effect of different preaging conditions on mechanical properties and clustering behavior of AA6016 sheets for automotive outer panels. *J. Mater. Res. Technol.* **2022**, *20*, 238–245.
11. Zhu, S.; Shih, H.C.; Cui, X.Y.; Yu, C.Y.; Ringer, S.P. Design of solute clustering during thermomechanical processing of AA6016 Al–Mg–Si alloy. *Acta Mater.* **2021**, *203*, 116455.
12. Chen, K.L.; Liu, C.H.; Ma, P.P.; Yang, J.S.; Zhan, L.H.; Huang, M.H.; Hu, J.H. Enhancing creep formability and comprehensive property in Al–Mg–Si alloy by combinatorial pre-ageing and large pre-deformation. *Mater. Sci. Eng. A* **2021**, *826*, 141967.
13. Chen, X.Y.; Guo, M.X.; Qiao, D.X.; Liu, L.K.; Yu, K.C.; Zhuang, L.Z. Simultaneously enhanced strength, ductility and formability of Al-Mg-Si-Cu-Zn-Fe-Mn alloy sheets by coupling distribution of solute, phases and grains. *Mater. Charact.* **2024**, *209*, 113720.

14. Lu, G.; Nie, S.; Wang, J.J.; Zhang, Y.; Wu, T.H.; Liu, Y.J.; Liu, C.M. Enhancing the bake-hardening responses of a pre-aged Al-Mg-Si alloy by trace Sn additions. *J. Mater. Sci. Technol.* **2020**, *40*, 107–112.
15. Zhang, M.X.; Wang, C.; Zhang, S.Y.; Liu, X.; Wang, X.; Ren, M.W.; Wang, H.Y. Enhanced aging precipitation behavior and mechanical properties of 6022 Al–Mg–Si alloy with Zr addition. *Mater. Sci. Eng. A* **2022**, *840*, 142957.
16. Li, G.J.; Guo, M.X.; Du, J.Q.; Zhuang, L.Z. Synergistic improvement in bake-hardening response and natural aging stability of Al-Mg-Si-Cu-Zn alloys via non-isothermal pre-aging treatment. *Mater. Des.* **2022**, *218*, 110714.
17. Xiao, Q.; Liu, H.Q.; Yi, D.Y.; Yin, D.Y.; Chen, Y.Q.; Zhang, Y.; Wang, B. Effect of Cu content on precipitation and age-hardening behavior in Al-Mg-Si-xCu alloys. *J. Alloys Compd.* **2017**, *695*, 1005–1013.
18. Zandbergen, M.W.; Cerezo, A.; Smith, G.D.W. Study of precipitation in Al–Mg–Si Alloys by atom probe tomography II. Influence of Cu additions. *Acta Mater.* **2015**, *101*, 149–158.
19. Thronsen, E.; Marioara, C.D.; Sunde, J.K.; Minakuchi, K.; Katsumi, K.; Erga, I.; Andersen, S.J.; Friis, J.; Marthinsen, K.; Matsuda, K.; et al. The effect of heavy deformation on the precipitation in an Al-1.3Cu-1.0Mg-0.4Si wt.% alloy. *Mater. Des.* **2020**, *186*, 108203.
20. Thronsen, E.; Mrkeseth, H.; Marioara, C.D.; Minakuchi, K.; Katsumi, T.; Marthinsen, K.; Matsuda, K.; Holmestad, R. The Effect of Small Additions of Fe and Heavy Deformation on the Precipitation in an Al–1.1Mg–0.5Cu–0.3Si At. Pct Alloy. *Metall. Mater. Trans. A* **2022**, *53*, 3296–3310.
21. Sunde, J.K.; Wenner, S.; Holmestad, R. In situ heating TEM observations of evolving nanoscale Al–Mg–Si–Cu precipitates. *J. Microsc.* **2019**, *279*, 143–147. [CrossRef] [PubMed]
22. Weng, Y.Y.; Jia, Z.H.; Ding, L.P.; Liao, J.; Zhang, P.P.; Xu, Y.Q.; Liu, Q. Effect of pre-straining on structure and formation mechanism of precipitates in Al–Mg–Si–Cu alloy. *Trans. Nonferrous Met. Soc. China* **2022**, *32*, 436–447.
23. Ding, L.P.; Jia, Z.H.; Liu, Y.U.; Weng, Y.Y.; Liu, Q. The influence of Cu addition and pre-straining on the natural aging and bake hardening response of Al-Mg-Si alloys. *J. Alloys Compd.* **2016**, *688*, 362–367.
24. Saito, T.; Muraishi, S.; Marioara, C.D.; Andersen, S.J.; Røyset, J.; Holmestad, R. The Effects of Low Cu Additions and Predeformation on the Precipitation in a 6060 Al-Mg-Si Alloy. *Metall. Mater. Trans. A* **2013**, *44*, 4124–4135.
25. *ASTM E8*; Standard Test Methods for Tension Testing of Metallic Materials. ASTM International: West Conshohocken, PA, USA, 2013.
26. Marioara, C.D.; Andersen, S.J.; Hell, C.; Frafjord, J.; Friis, J.; Bjørge, R.; Ringdalen, I.G.; Engler, O.; Holmestad, R. Atomic structure of clusters and GP-zones in an Al-Mg-Si alloy. *Acta Mater.* **2024**, *269*, 119811.
27. Martinsen, F.A.; Ehlers, F.J.H.; Torsæter, M.; Holmestad, R. Reversal of the negative natural aging effect in Al–Mg–Si alloys. *Acta Mater.* **2012**, *60*, 6091–6101.
28. Werinos, M.; Antrekowitsch, H.; Ebner, T.; Prillhofer, R.; Curtin, W.A.; Uggowitzer, P.J.; Pogatscher, S. Design strategy for controlled natural aging in Al–Mg–Si alloys. *Acta Mater.* **2016**, *118*, 296–305.
29. Li, Y.F.; Wu, Z.B.; Wang, D.T.; Nagaumi, H.; Zhang, G.Y.; Feng, Z.X.; Luo, X.H.; Luo, Y.Z.; Zhang, H.T.; Zhamg, B. Synergistic enhancement of strength and bendability of Al-Mg-Si-Cu alloys for safety components by adding Mn. *Mater. Des.* **2025**, *249*, 113531.
30. Saito, T.; Mrtsell, E.A.; Wenner, S.; Marioara, C.D.; Andersen, S.J.; Friis, J.; Matsuda, K.; Holmestad, R. Atomic structures of precipitates in Al–Mg–Si alloys with small additions of other elements. *Adv. Eng. Mater.* **2018**, *20*, 1800128.
31. Cao, L.F.; Rometsch, P.A.; Couper, M.J. Clustering behaviour in an Al–Mg–Si–Cu alloy during natural ageing and subsequent under-ageing. *Mater. Sci. Eng. A* **2013**, *559*, 257–261.
32. Mantina, M.; Wang, Y.; Chen, L.Q.; Liu, Z.K.; Wolverton, C. First principles impurity diffusion coefficients. *Acta Mater.* **2009**, *57*, 4102–4108. [CrossRef]
33. Tu, W.B.; Tang, J.G.; Zhang, Y.; Ye, L.Y.; Liu, S.D.; Lu, J.K.; Zhan, X.; Li, C.B. Effect of Sn and Cu addition on the precipitation and hardening behavior of Al-1.0Mg-0.6Si alloy. *Mater. Sci. Eng. A* **2020**, *770*, 138515. [CrossRef]
34. Cui, M.; Lim, S.H.; Kim, H.; Lee, J. The effect of Cu additions and prestraining on the age-hardening behavior of AA6016 sheets. *J. Alloys Compd.* **2023**, *960*, 170801. [CrossRef]
35. Serizawa, A.; Sato, T.; Miller, M.K. Effect of cold rolling on the formation and distribution of nanoclusters during pre-aging in an Al–Mg–Si alloy. *Mater. Sci. Eng. A* **2013**, *561*, 492–497. [CrossRef]
36. Ninive, P.H.; Strandlie, A.; Gulbrandsen-Dahl, S.; Lefebvre, W.; Marioara, C.D.; Andersen, S.J.; Friis, J.; Holmestad, R.; Løvvik, O.M. Detailed atomistic insight into the β'' phase in Al–Mg–Si alloys. *Acta Mater.* **2014**, *69*, 126–134. [CrossRef]

Disclaimer/Publisher's Note: The statements, opinions and data contained in all publications are solely those of the individual author(s) and contributor(s) and not of MDPI and/or the editor(s). MDPI and/or the editor(s) disclaim responsibility for any injury to people or property resulting from any ideas, methods, instructions or products referred to in the content.

materials

MDPI

Article

Analysis of the Wear and Corrosion Resistance on Cu-Ni-Al Composites Reinforced with CeO_2 Nanoparticles

Carola Martínez [1,*], Bárbara Valverde [2,*], Aurora Del Valle-Rodríguez [3], Brennie Bustos-De La Fuente [1], Izabel Fernanda Machado [4] and Francisco Briones [2]

1 Departamento de Ingeniería en Obras Civiles, Universidad de La Frontera, Temuco 4811230, Chile; b.bustos05@ufromail.cl
2 Escuela de Ingeniería Mecánica, Facultad de Ingeniería, Pontificia Universidad Católica de Valparaíso, Quilpué 2430120, Chile; francisco.briones@pucv.cl
3 Programa de Magister en Ciencias de la Ingeniería, Universidad de La Frontera, Temuco 4811230, Chile; aurora.delvalle@ufrontera.cl
4 Departamento de Engenharia Mecatrônica e Sistemas Mecânicos, Escola Politecnica, Universidade de São Paulo, São Paulo 05508-030, Brazil; machadoi@usp.br
* Correspondence: carola.martinez@ufrontera.cl (C.M.); barbara.valverde@pucv.cl (B.V.); Tel.: +56-452596818 (C.M.); +56-937109993 (B.V.)

Abstract: This study evaluates the wear and corrosion resistance of the Cu-50Ni-5Al alloy reinforced with CeO_2 nanoparticles for potential use as anodes in molten carbonate fuel cells (MCFCs). Cu–50Ni–5Al alloys were synthesized, with and without the incorporation of 1% CeO_2 nanoparticles, by the mechanical alloying method and spark plasma sintering (SPS). The samples were evaluated using a single scratch test with a cone-spherical diamond indenter under progressive normal loading conditions. A non-contact 3D surface profiler characterized the scratched surfaces to support the analysis. Progressive loading tests indicated a reduction of up to 50% in COF with 1% NPs, with specific values drop-ping from 0.48 in the unreinforced alloy to 0.25 in the CeO_2-doped composite at 15 N of applied load. Furthermore, the introduction of CeO_2 decreased scratch depths by 25%, indicating enhanced wear resistance. The electrochemical behavior of the samples was evaluated by electrochemical impedance spectroscopy (EIS) in a molten carbonate medium under a H_2/N_2 atmosphere at 550 °C for 120 h. Subsequently, the corrosion products were characterized using X-ray diffraction (XRD), scanning electron microscopy coupled with energy dispersive spectroscopy (SEM-EDS), and X-ray photoelectron spectroscopy (XPS). The results demonstrated that the CeO_2-reinforced alloy exhibits superior electro-chemical stability in molten carbonate environments (Li_2CO_3-K_2CO_3) under an H_2/N_2 atmosphere at 550 °C for 120 h. A marked reduction in polarization resistance and a pronounced re-passivation effect were observed, suggesting enhanced anodic protection. This effect is attributed to the formation of aluminum and copper oxides in both compositions, together with the appearance of NiO as the predominant phase in the materials reinforced with nanoparticles in a hydrogen-reducing atmosphere. The addition of CeO_2 nanoparticles significantly improves wear resistance and corrosion performance. Recognizing this effect is vital for creating strategies to enhance the material's durability in challenging environments like MCFC.

Keywords: Cu-Ni-Al alloys; CeO_2 nanoparticles; scratch test; corrosion

1. Introduction

Fuel cells (FCs) come in various types, differing in the electrolyte used, operating temperatures, and applications. The FCs are classified according to the operation temperature, the chemical nature of the electrolyte, and the fuel type [1]. The high-temperature technologies for generating electrical energy are molten carbonate fuel cells (MCFCs) and solid oxide fuel cells (SOFCs) [2] due to their high-power densities and efficiencies (100 MW) for stationary applications as power generators [3,4]. However, the MCFC operates at lower temperatures than SOFC, close to 650 °C, with a higher life span of 7000 h (SOFC 1000 h) [5], with flexibility of fuel, a low investment catalyst, and low cost [6]. This is why it is considered an attractive option to develop medium-scale stationary units that generate between 100 kW and 100 MW [7]. MCFC comprises an anode and a cathode separated by a molten carbonate electrolyte. The primary issues with MCFC are the stability and degradation of materials caused by high temperatures, which decrease cell efficiency [8].

Ni-Al anodes are widely used in MCFC applications due to their relatively low density, high melting point, excellent corrosion resistance to acids and alkalis, and good charge transfer conductivity [9]. Anode materials are typically Ni–Cr/Ni–Al alloys [10,11], and cathode materials are comprised of lithiated NiO [12]. The fuel cell operation results from a complex conjunction of physical, chemical, and electrochemical processes. The anode and cathode half reactions and overall electrochemical reaction are [13,14]:

$$\text{Cathode: } \tfrac{1}{2}\,O_2 + CO_2 + 2e^- \rightarrow CO_3^{2-}$$

$$\text{Anode: } H_2 + CO_3^{2-} \rightarrow H_2O + CO_2 + 2\bar{e}$$

$$\text{Overall: } H_2 + \tfrac{1}{2}\,O_2 + CO_{2\,(cathode)} \rightarrow H_2O + CO_{2\,(anode)} + \text{electricity} + \text{heat}$$

Due to the high temperature, they are susceptible to creep, which causes a decrease in sintering resistance and a reduction in electrode pore size, dramatically shortening the lifespan of fuel cells [9,15]. In this context, generating efficient and stable anodes by incorporating alloy additives requires understanding the underlying mechanisms to design other efficient and stable anodes. For this reason, incorporating other metals, such as copper, was investigated to enhance structural stability [16,17].

The incorporation of copper (Cu) into the Ni-Al alloy presents an opportunity to enhance thermal conductivity and mechanical strength over a wide range of temperatures, making it an ideal material for operating under extreme conditions of heat flow and compression [10,17]. Copper is less expensive and more stable under accidental current overload than nickel because of its positive potential. Furthermore, copper has higher electrical conductivity and greater resistance to carburization than nickel [16]. According to Klassert and Tikana [11], copper alloys are an important metallic material due to their metallurgical, physical, and chemical properties. For example, Cu-Ni alloys show high tensile strength and good corrosion resistance at high temperatures. Other authors, such as Martínez et al. [18], study Cu and Ni separately, as well as the Cu-Ni alloy, indicating that Ni exhibits the maximum resistance. However, the Cu-Ni alloy has intermediate values between Cu and pure Ni. It is essential to mention that the incorporation of metal can stabilize the alloy in a solid solution, which means that solvent metal atoms are randomly replaced by solute metal atoms in their positions in the lattice, creating metallic bonds [19,20].

Alternatively, the incorporation of nanoparticles in commercial alloys, such as Ni-5Al, has been studied to enhance the mechanical behavior of the anode, incorporating zirconia oxides (ZrO_2) [21,22] and cerium oxides (CeO_2) [23]. The addition of nanoparticles to the Ni-5Al alloy reduces creep deformation by enhancing microstructural stability, catalytic activity, and cell efficiency [18]. Some authors have reported that incorporating CeO_2 nanoparticles enhances the catalyst performance at the anode for the H_2 oxidation reaction [22,23]. Additionally, incorporating cerium oxide (CeO_2) nanoparticles into Cu-Ni alloys has improved the catalyst performance in the hydrogen oxidation reaction [24]. Recently, studies have focused on the effects of hydrogen atmospheres on the durability and efficiency of fuel cell components. For example, research by Accardo et al. [22] demonstrated that exposure to hydrogen can lead to the formation of nickel hydrides on the anode surface, affecting its electrochemical properties. Furthermore, hydrogen embrittlement is a phenomenon where materials become brittle due to hydrogen absorption, posing a significant challenge to the structural integrity of fuel cell components [22]. However, further research is still required on the micro- and macrostructural behavior and corrosion resistance of these nanoparticle-reinforced alloys in hydrogen atmospheres.

On the other hand, there are few studies on the tribological behavior of the components used in a fuel cell (FC). Although MCFC anodes are not subject to continuous mechanical friction during operation, their surfaces may suffer localized damage during stack assembly or thermal cycling. Scratch resistance is an indicator of surface hardness and mechanical stability, which helps preserve the anode's porosity and functionality. Recent studies have shown that enhancing wear resistance in functional surfaces, even without active sliding, contributes to improved durability and electrochemical performance [25,26]. Venkatesh et al. [27] evaluate friction wear damage in a single microcontact at different contact pressures on a proton-exchange membrane fuel cell. However, microcontacts are likely to occur during service in the fuel cell assembly, which can affect fuel cell performance. V. E. Pukha et al. [28] investigate CNC-coated titanium as a viable alternative to gold-coated stainless steel for bipolar plates in proton exchange membrane fuel cells, focusing on its corrosion, low interface contact resistance (ICR), and high wear resistance. However, it is expected that CeO_2 nanoparticles, due to their lubricating properties [29], could decrease the coefficient of friction, thereby improving tribological performance by reducing wear and increasing the useful life of the anodes. The above is considering that the addition of CeO_2 nanoparticles to copper-based alloys can simultaneously enhance their mechanical, tribological, and thermal stability properties. At the microstructural level, CeO_2 acts as a grain refiner and a barrier to dislocation motion, contributing to increased hardness and improved high-temperature creep resistance [30]. From a tribological perspective, nanoscratch and wear studies have demonstrated that CeO_2 facilitates the formation of stable and homogeneous tribochemical films, reducing friction and protecting the surface from progressive wear [31]. In addition, in spark plasma sintered composites such as CeO_2-doped Cu/WS_2, these nanoparticles help control interfacial microstructure and prevent thermal degradation during processing, leading to materials with high densification, low friction, and good mechanical strength [32].

Therefore, this work aims to analyze the influence of CeO_2-NPs as a reinforcing material in Cu-Ni-Al alloys in the singular scratch test and the corrosion resistance in molten salt of carbonate in an H_2/N_2 atmosphere, in order to analyse their possible use as an anode in a molten carbonate fuel cell (MCFC).

2. Materials and Methods

2.1. *Samples Manufacturing*

Sample manufacturing was performed using mechanical alloying and mechanical mixing. Mechanical alloying was performed with pure powders: Cu (99.7% at., <63 mesh, Merck, Darmstadt, Germany), Ni (99% at., <230 mesh, Merck), and Al (<60 μm, 99.9%, Good Fellow, Peterlee, UK). The mechanical alloying was carried out with a powder ratio of Cu-50Ni-5Al (wt.%) in a planetary mill (Across International model PQ-N04, Bayswater,

VIC, Australia) using stainless steel containers and balls (ball diameters: 10 and 6 mm). The containers were filled in a glove box under an argon atmosphere. The ball-to-powder ratio (BPR) was 10:1, and 2 wt.% stearic acid was used as a control agent. The angular velocity is 350 r.p.m., and a cycle of 30/15 min on/off is used to maintain a constant temperature over a 30 h milling time.

The 1% CeO_2 nanoparticles (CeO_2-NPs) were incorporated in Cu-50Ni-5Al in a mixed Y–type Astecma for 1 h. Subsequently, Cu-50Ni-5Al samples without and with CeO_2-NPs were consolidated by Spark Plasma Sintering (SPS) using a Fuji Electronic Industrial Co. model DR SINTER SPS1050 (Tsurugashima, Japan), according to the procedure previously described by Martínez et al. [18]. The samples obtained have a 16 ± 0.8% porosity for Cu-50Ni-5Al and 1.0 ± 0.3% for Cu-50Ni-5Al + 1 wt.% CeO_2-NPs; and the hardness values of 0% CeO_2–NPs is 205 ± 21 $HV_{0.3}$ and 1% CeO_2–NPs is 340 ± 37 $HV_{0.3}$ [18].

A Mecatome–Presi machine mechanized Cu-Ni-Al and Cu-Ni-Al + 1% CeO_2-NPs samples. Then, a Dremel was used to make a hole in the top of the samples for electrochemical analysis.

2.2. Singular Scratch Test

The samples were metallographically prepared by reducing the grit size of the abrasive from #800, #1000, to #1200 SiC sandpaper. Subsequently, automatic polishing was performed using a Struers polisher, model Labopol-60 (Champigny-sur-Marne, France), with cloths for the different suspensions of diamond particles (diamond paste) of 6, 3, and 1 µm. To obtain the final surface, the sample was polished on colloidal silica (0.04 µm). Finally, to eliminate particles on the sample surfaces, they were left in alcohol under ultrasound and dried in air.

Scratch tests at this scale were used with a Bruker UMT-2 Tribometer (Billerica, MA, USA). The scratching procedures were performed using a Rockwell-C spherical diamond indenter (Milwaukee, WI, USA) with a radius of 200 µm and a 120° spherical radius—the test protocol employed a progressive loading Scratch Test. The normal load was increased linearly from 5 to 35 N during the test. The coefficient of friction is the ratio of the tangential force (F_t) to the normal force (F_n). The curve plotted with different loads is an average of three scratch measurements. The profiles of wear tracks were analysed using the 3D interferometry technique with the Taylor-Hobson model CCI MP equipment (Leicester, UK). Each scratch was inspected at three positions between the starting and end points to measure the pile-up area (Force normal: 7 N, 23 N and 33 N), scratch area, and scratch depth. The CCI-MP performs area measurements on the mean profile of the scratch section appearing in the field of view and provides the average values. The data obtained were analyzed using the software Talymap Gold v6.2.661.

2.3. Electrochemical Measurements

The electrochemical behavior of the Cu-50Ni-5Al samples was studied using open circuit potential and electrochemical impedance spectroscopy (EIS) measurements in the molten Li_2CO_3-K_2CO_3 (62:38 mol. %) as electrolyte at 550 ± 5 °C under a controlled media were performed using a constant flow of H_2/N_2 in a 40/10 mL·min^{-1} ratio. The electric contact for the working electrode, the Cu-50Ni-5Al samples, was performed using conductive silver printing ink (resistivity 5–6 µΩ cm) around the sample and copper wire of 25 cm in length. In addition, a Pt wire was used as the counter electrode, measuring 25 cm in length. An Ag wire, also 25 cm in length, was placed inside a quartz glass tube with a porous plug at the tip, serving as the reference electrode. The electrochemical measurements were performed using a Potentiostat from Solartron Analytical (Farnborough, UK).

Figure 1 provides a summary diagram illustrating the methodology employed in this study.

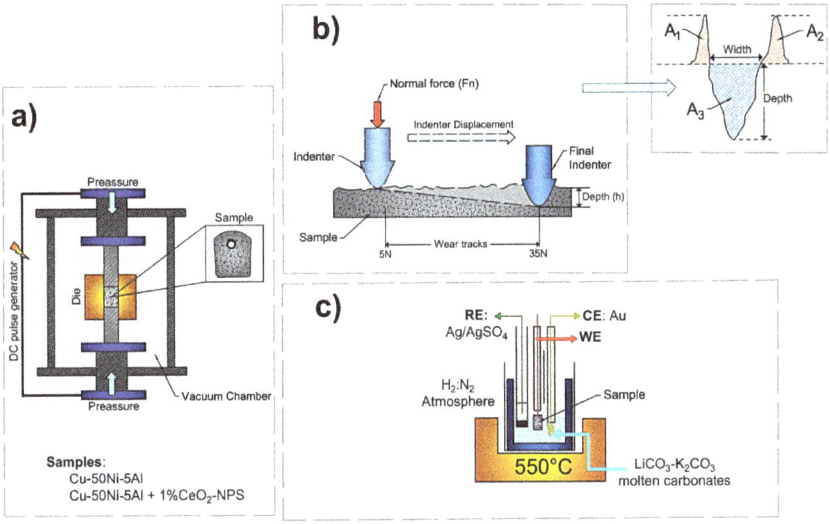

Figure 1. Schematic representation of the methodology: (**a**) SPS system (consolidation), (**b**) single scratch test, and (**c**) set-up of half-cell to Li_2CO_3-K_2CO_3 at 550 °C under an H_2/N_2 atmosphere.

2.4. Morphological and Microstructural Analysis

Morphological characterization was performed using scanning electron microscopy (SEM) with a Hitachi SU3500 microscope (Tokyo, Japan) equipped with an EDX analyser from Bruker XFlash, 15 kV accelerating voltage, and 155.000 nA current. The samples were characterized by X-ray diffraction (XRD) to identify phases and assess phase purity. X-ray powder diffraction patterns were obtained using a powder diffractometer Bruker D2 Phase equipped with a Cu-K_α radiation source. The diffraction patterns were recorded with a 2θ range of 20° to 100°, a step size of 0.02°, and a counting time of 1 s per step. Diffraction patterns were analyzed using MACH! 3 Software version 3.15 build 258 to identify the phases present. X-ray photoelectron spectroscopy (XPS) was also performed to study the surface composition in more detail, utilizing an analysis chamber, Phoibos 150 (SPECS Group, Berlin, Germany), equipped with a 1D-DLD detector and a monochromatic X-ray source: Focus 500 with an XR50 M, featuring an Ag/Al Anode and X-ray source power of 400 W. The XPS data were analyzed using Casa XPS Version 2.3.22 software, calibrating the energy scale of the spectra with the binding energy of the C1s signal at 285.4 eV, which allowed us to remove the Shirley-type background. A mixed Gauss–Lorentz shape was used for the different components.

3. Results

3.1. Scratch Test with Progressive Normal Load

Figure 2 presents the scratch test results for Cu-50Ni-5Al + 0% CeO_2-NPs (0% CeO_2-NPs) and Cu-50Ni-5Al + 1% CeO_2-NPs (1% CeO_2-NPs) samples. Figure 2a shows the coefficient of friction (COF) as a function of normal force (Fn). The average coefficient of friction (COF) values obtained for the three normal loads evaluated show consistent differences between the samples with and without the addition of CeO_2-NPs. In the unreinforced alloy, values of 0.24 ± 0.06 (7 N), 0.49 ± 0.02 (23 N), and 0.49 ± 0.02 (33 N) were recorded, while in the samples with 1% CeO_2, values of 0.22 ± 0.04 (7 N), 0.46 ± 0.03

(23 N), and 0.48 ± 0.01 (33 N) were obtained. The observed differences remain consistent at all load levels. It is important to highlight that the dispersion levels are within the ranges typically reported for metallic materials evaluated by scratch tests, as pointed out in previous studies [33].

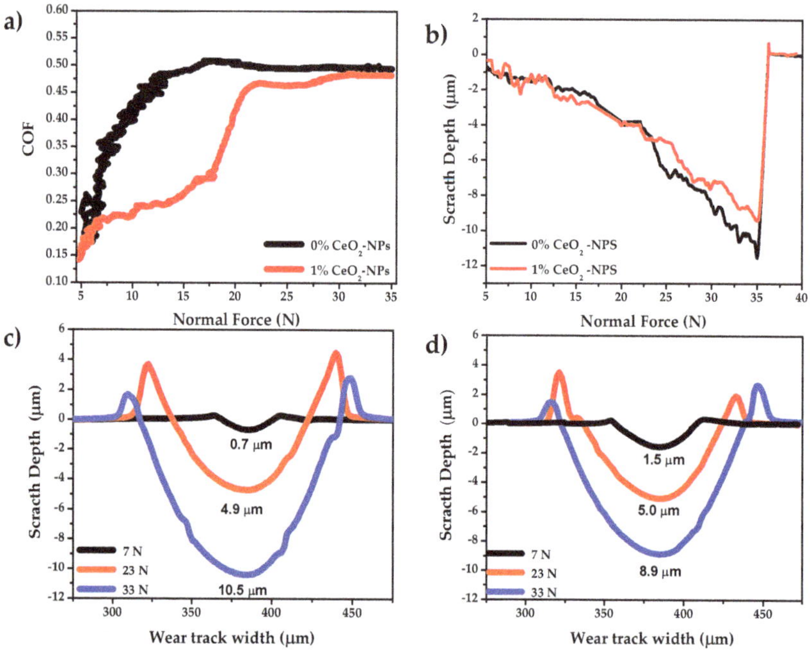

Figure 2. (**a**) COF vs. normal Force, (**b**) normal force vs. scratch depth, and the wear tracks profile (**c**) 0% CeO_2–NPs, and (**d**) 1% CeO_2-NPs.

The 0% CeO_2-NPs sample exhibits more significant COF oscillations up to approximately 13 N. Then, the friction curve increases and stabilizes at approximately 21 N and remains without significant changes in the slope as the force reaches the maximum load of 35 N.

For the 1% CeO_2-NPs, small oscillations are observed in the coefficient of friction (COF) curve up to approximately 7 N. After this point, there is a significant change in the slope of the curve that extends to 17 N, followed by another increase in the slope up to 22 N. This behavior may be associated with a transition region characterized by plastic deformations occurring below, besides, and in front of the tip [34]. Around 23 N, the COF curve shows no significant changes in slope, resulting in the same COF values for the 0% CeO_2-NPs sample. This phenomenon is attributed to the material removal, which alters the tangential force proportionately to the normal force. It is important to note that between 7 N and 23 N, there are significant differences in the COF between the two samples. Specifically, the sample containing 1% CeO_2-NPs shows a remarkable reduction in COF of more than 50%, reaching a value of 0.25 ± 0.01 at 15 N. In contrast, the 0% CeO_2-NPs sample has a COF of 0.48 ± 0.02. This decrease in COF can be attributed to the presence of load-bearing nanoparticles, which reduce the actual contact area and the interaction of asperities between the contacting surfaces. This interaction varies due to differences in hardness between the samples and the presence of pores during the sliding of the indenter on the surface where the indenter penetrates deeper, increasing the friction force [35] (0% CeO_2-NPs: 16 ± 0.8% de porosity; 1% CeO_2-NPs: 1.0 ± 0.3% porosity [18]).

Figure 2b shows the scratch depth curves as a function of normal load. Starting at 23 N, the results indicate a separation between the two curves, reaching an average difference between depths of 25%. Due to these transitions, three normal loads, specifically 7, 23, and 33 N, were selected for further analysis.

Figure 2c and 2d shows the profile of the wear track and the pile-up of both samples, indicating that when the indenter (hard particle) scratches the alloy containing 1% CeO_2-NPs, it produces a shallower scratch. These results indicate that the sample with 0% CeO_2-NPs will experience greater mass loss than 1% CeO_2-NPs. It is important to note that higher hardness is directly correlated with shallower scratch depths, which aligns with the general principle that harder materials exhibit better resistance to wear and abrasion [36] (0% CeO_2–NPs: 205 ± 21 $HV_{0.3}$ and 1% CeO_2–NPs: 340 ± 37 $HV_{0.3}$ [18]). This observation is consistent with widely reported findings in the literature showing that increased surface hardness reduces plastic deformation and the material removal rate during contact, which ultimately improves wear resistance. Numerous metallic and composite systems studies support this correlation [37–39]. Furthermore, applying higher normal loads is expected to increase the scratch depth. On a microscale, this interaction varies within a specific range throughout the test period due to hardness differences between samples and the presence of pores during the sliding of the indenter on the surface [35].

Another important aspect is the accumulation of material along the edges of the scratch track, which is evident in all samples under different applied loads. This accumulation of displaced material indicates significant plastic deformation during the tip-sliding process. The pile-up formation suggests the material exhibits a relatively low strain-hardening capacity, allowing localized plastic flow rather than extensive fracture or substantial material removal. These effects were more pronounced at a load of 23 N in both cases, which is attributed to the fact that the magnitude of pile-up is influenced by both the applied load and the material hardness [40]. The more pronounced pile-up formation in the 0% CeO_2–NPs sample could be attributed to its higher porosity, which reduces the local hardness and favors material displacement rather than fragmentation. In contrast, the lower stacking height in the 1% CeO_2–NPs sample is probably due to the effect of the nanoparticles, which contributes to higher hardness, resistance to localized deformation, and lower presence of pores during tip sliding.

Table 1 presents the wear profiles' main area and dimension parameters for the three selected loads. The first ratio, A_1+A_2, is associated with the pile-up, and the A_3 value is associated with the material that originally occupied the tracks. Attributing its destination to the formation of piles and wedges or effective removal by micro-cutting is unnecessary. These area ratios are justified by considering practical situations in which the portion of material that has moved out of the tracks (pile and wedge) can be removed with increases in normal force. Therefore, the ratio of these areas can suggest predominant micro-mechanisms or transitional micro-mechanisms [41], where it is observed that relationships close to 0 indicate micro-cutting as the predominant micro-mechanism (values of 33 N), and values very close to 1 indicate micro-ploughing as dominant at low loads (7 N). Between these values, there would be a transition of micro-mechanisms. Additionally, the main differences between the results in A_1+A_2 in the profiles at 23 N could be associated with the fact that the undetected chips cause an increase in the measurements of the stacking areas, where Franco et al. [42] address these false accumulations, which is a delicate point in area measurement.

Figure 3 shows SEM images of both samples between 6.5 to 10 N of applied load. Generally, during the indenter passage, most of the material appears to sink to the bottom of the tracks, and a small amount shifts to the edges, forming the pile-up. This micro-mechanism is well known in the literature as micro-ploughing [43]. Figure 3c shows the wear track of the 0% CeO_2-NPs sample at higher magnification. The figure shows the presence of pores, in which the loading conditions may tend to cause the closure of surface pores during tip sliding, which influences how the material deforms during the test [44]. The tendency towards pore closure may explain the differences in the geometric parameters of the wear profile between both samples at 7 N (see Table 1), causing the increase in contact pressure that promoted high plastic deformation around the surface pores. This localized deformation induced the surface pores to collapse instead of moving towards the edges and bottom of the tracks [44,45].

Table 1. Parameters of the dimensions and areas from the wear profiles for the three selected loads.

Sample	Load (N)	Depth (μm)	Width (μm)	A_1+A_2 (μm^2)	A_3 (μm^2)	Pile-Up Ratio
0% CeO_2-NPs	7	0.7 ± 0.10	31.6	9.7	13.9	0.70
	23	4.9 ± 0.06	84.3	111.5	277.0	0.42
	33	10.5 ± 0.02	124.7	45.3	842.0	0.05
1% CeO_2-NPs	7	1.5 ± 0.12	50.8	7.5	50.9	0.15
	23	5.0 ± 0.04	87.7	53.23	280.3	0.18
	33	8.9 ± 0.03	116.2	49.5	673.8	0.07

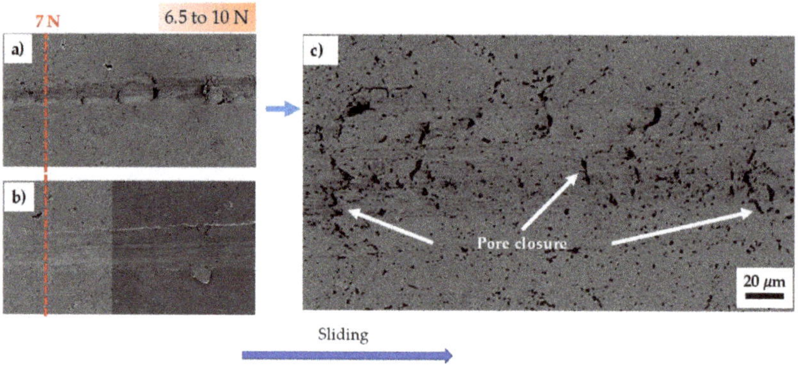

Figure 3. Wear path of the (**a**) 0% CeO_2-NPs, (**b**) 1% CeO_2-NPs, and (**c**) zoom in the wear track 0% CeO_2-NPs. Applied normal force: between 6.5 and 10 N.

Figure 4 shows the wear path for distinct applied normal force. At 16 to 25 N normal force (Figure 4a,b), the track becomes wider and exhibits more intense plastic deformation within the track. A micro-mechanism transition can be seen from micro-ploughing to micro-cutting, where the accumulation of material on the edges generates the formation of chips as the normal load increases. This micro-mechanism of transition from mild to severe abrasion often occurs similarly in both materials. The transition is attributed to the fact that in ductile materials, it is common to find chips that do not detach from the surface and remain near the edge of the track [42], as seen in both SEM images, which supports the findings presented in Table 1.

At loads of 28 and 35 N, a more pronounced plastic deformation is observed within the tracks (see Figure 4c,d). In addition, it is seen that there are regions with more intense interaction between the indenter and the surface, typical of micro-cutting: chip formation, chip detachment, and chips next to the track. Both tracks show significant width variations and regions with poorly defined edges. This could explain the differences in parameter A_3 between both samples, which show significant differences; the 1% CeO_2-NPs sample has a 20% smaller area than the 0% CeO_2-NPs sample. The use of this parameter is justified considering practical situations in which the portion of the material that has moved outside the track is easily removed at those load levels (33 N). As a result, the 0% CeO_2-NPs sample shows more material displaced outside the track and greater penetration depths (Figure 2). These values indicate the advantages of using nanoparticles at high normal force.

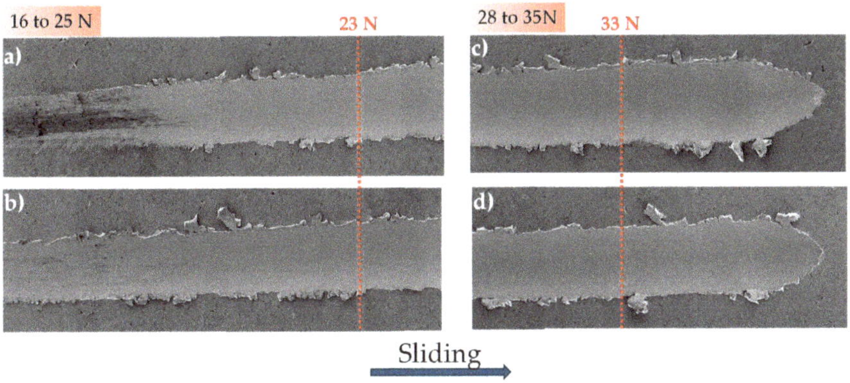

Figure 4. Wear path of the (**a**,**c**) 0% CeO_2-NPs, (**b**,**d**) 1% CeO_2-NPs. Applied normal force: between 16 to 25 N and 28 to 35 N.

3.2. Electrochemical Measurements

Figure 5 shows the open circuit potential (E_{OC}) of the effect of 0% CeO_2-NPs and 1% CeO_2-NPs after exposure to Li_2CO_3-K_2CO_3 at 550 °C in H_2/N_2 atmosphere. After a shorter exposure time, the E_{OC} shifted to more negative values upon incorporating CeO_2-NPs, suggesting an activation of corrosion phenomena. However, for longer exposure times, the E_{OC} reached potential values above the sample without CeO_2-NPs of approximately −0.43 V vs Ag/Ag+, which can be attributed to a stable oxide layer formed on the metal surface [46]. The 0% CeO_2-NPs sample exhibits a decrease in the E_{OC} at shorter exposure times; however, after 20 h, relatively constant values are observed. This behavior can be attributed to Al, forming a protective alumina layer on the surface, preventing continued reaction at higher temperatures [47].

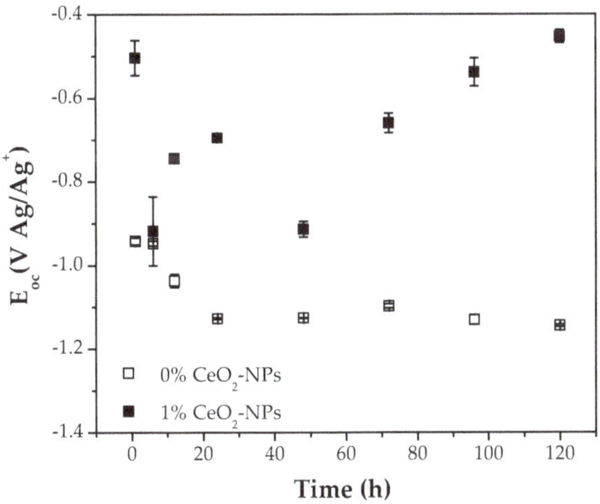

Figure 5. Open circuit potential variation of (□) 0% CeO_2-NPs and (■) 1% CeO_2-NPs during exposure to Li_2CO_3-K_2CO_3 at 550 °C under an H_2 atmosphere.

Figure 6 shows the Nyquist diagrams of 0% CeO_2-NPs and 1% CeO_2-NPs samples after 120 h of exposure to Li_2CO_3-K_2CO_3 in a H_2 atmosphere at $E = E_{OC}$ and 550 °C. The impedance analysis reveals a capacitive behavior of the alloys, independent of CeO_2-NPs, indicating a higher resistance to polarization as a function of increased exposure time for both samples, attributed to the formation of protective oxides [18]. Concerning the effect of temperature on the reaction rate, the anodic reaction has positive activation energies [48], indicating that the oxidation of H_2 is facilitated at higher temperatures. The heat of the reaction for the gaseous dissolution of H_2 and CO_2 in carbonates also has positive values [49]. Therefore, a higher temperature increases the rate of gas dissolution.

$$H_2 + CO_3{}^{-2} \rightarrow CO_2 + H_2O + 2\bar{e} \quad (1)$$

$$\tfrac{1}{2}\, O_2 + CO_2 + 2\bar{e} \rightarrow CO_3{}^{-2} \quad (2)$$

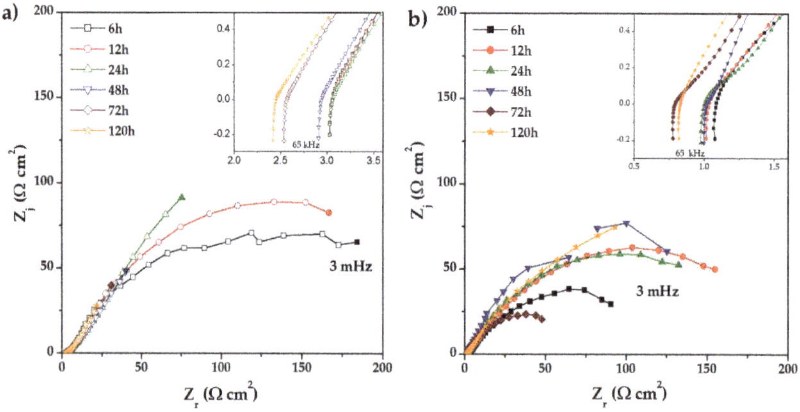

Figure 6. Nyquist diagram of (**a**) 0% CeO_2-NPs, and (**b**) 1% CeO_2 NPs during exposure to Li_2CO_3-K_2CO_3 at 550 °C under an H_2/N_2 atmosphere.

The incorporation of 1% CeO_2-NPs decreases the resistance to polarization, which is observed in a less capacitive arc at all exposure times; this may be associated with the protective properties of the oxide layer formed on the surface, such as its poor adhesion and heterogeneity [50]. It should be noted that at 120 h, the 1% CeO_2-NPs sample presents better behavior than the 0% CeO_2-NPs sample. This can be attributed to the difference in the corrosion products formed between the two systems, which reduces electrochemical activity and cell performance.

As can be observed from $E = E_{OC}$, the impedance responses reveal two-time constants in the high- and low-frequency ranges (HF and LF), which can be associated with the cathodic current. This current involves the capacitance of the electric double layer (C_{dl}) and the formation of an oxide layer due to the dissolution of the alloy, which is related to the surface's heterogeneity [51].

Figure 7 shows the Bode diagrams of 0% CeO_2-NPs and 1% CeO_2-NPs samples after 120 h of exposure to Li_2CO_3-K_2CO_3 in an atmosphere of H_2 at $E = E_{OC}$ and 550 °C. The modulus impedance diagrams did not reveal significant differences in the LF range when CeO_2-NPs were added. Note that in the sample with nanoparticles, the low-frequency impedance module decreases to 56.5 Ω cm^2 after 72 h of exposure and then increases to 120.7 Ω cm^2 after 120 h of exposure. This is likely associated with the oxide film cracks and then becomes passivated after 72 h. Figure 7b,d also show the phase angles for both samples. These do not present significant differences but reveal a capacitive response with two- or three-time constants in all frequency ranges, possibly associated with forming an oxide film [52].

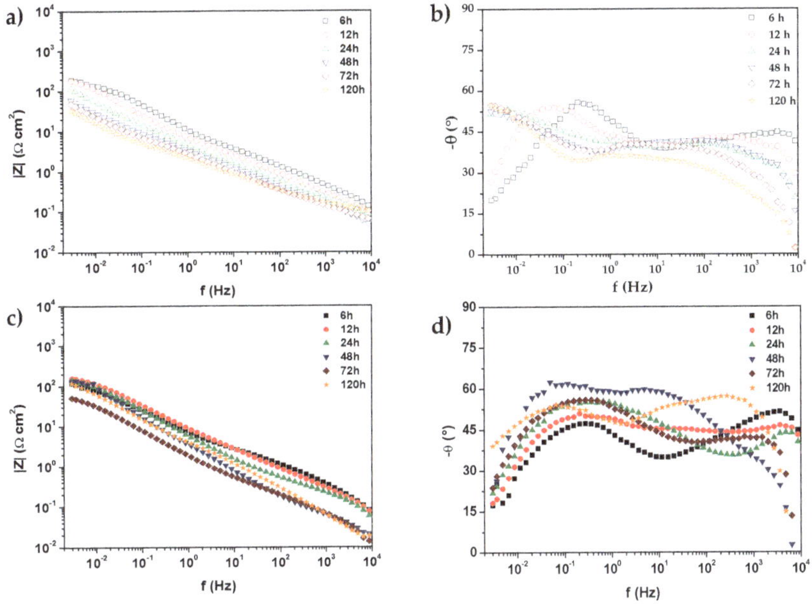

Figure 7. Corrected Bode plots of (**a**,**b**) 0% CeO_2-NPs, and (**c**,**d**) 1% CeO_2-NPs in Li_2CO_3-K_2CO_3 at 550 °C under an H_2/N_2 atmosphere.

Angles close to 20° are revealed at short exposure times at LF, 50° at longer exposure times at LF, and angles between 35 and 50° at HF. Due to the geometry of the electrode and surface heterogeneity, this response may be attributed to frequency dispersion [50]. Lee [53] studied the influence of temperature on the anode reaction in a molten carbonate fuel cell. Their results revealed that the high-frequency (HF) circle shrinks as the temperature increases, but the low-frequency (LF) circle increases. According to Lee [54], HFC represents cathodic mass transfer resistance through the electrolyte film, whereas LF represents gas-phase mass transfer resistance, mainly due to anode gas flow. Since the anode flow has a lower Reynolds number than the cathode, the anode has a much thicker boundary layer. The thickness of the layer causes mass transfer resistance in the gas phase. Therefore, the reduced size of HF indicates that the cathodic resistance decreases with increasing temperature. This is acceptable because an increase in temperature increases the solubility and diffusivity of the gas in the electrolyte film, thereby reducing mass transfer resistance.

Figure 8 shows the variation of the imaginary part of the impedance of the samples with and without nanoparticles as a function of frequency after 120 h of exposure to Li_2CO_3-K_2CO_3 in a H_2/N_2 atmosphere at $E = E_{OC}$ and 550 °C, revealing a constant phase element (CPE) behavior in the MF range, with a negative slope (α) varying between -0.45 ± 0.003 for 0% CeO_2-NPs and -0.40 ± 0.04 for 1% CeO_2-NPs, which can be related to the oxide film formed on the alloy surface (Z_{oxide}) and described by the following relationship, as reported by Orazem and Tribollet [55], Tribollet et al. [56], and Hirschorm et al. [57].

$$Z_{oxide} = \int_0^{\delta} \frac{\rho(\gamma)}{1 + j\omega\rho(\gamma)\varepsilon(\gamma)\varepsilon_0} d\gamma \quad (3)$$

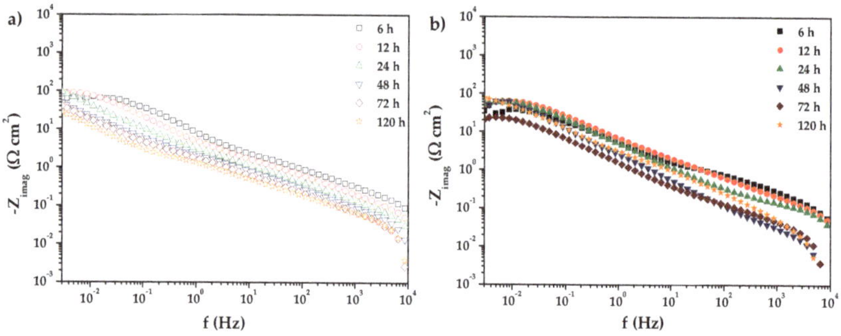

Figure 8. Log Z_{imag} of (**a**) 0% CeO_2-NPs and (**b**) 1% CeO_2–NPs in Li_2CO_3-K_2CO_3 at 550 °C under an H_2/N_2 atmosphere.

The authors proposed the power law model (PLM) to analyze the physical properties of the film as described by the Equation (4):

$$Z(\omega) = g\frac{\delta\rho_{\delta}^{1-a}}{(\rho_0^{-1} + j\omega\varepsilon\epsilon_0)^a} \quad (4)$$

In this case, α is the slope in the Log Z_{Imag} vs Log f graphs, ε represents the dielectric constant of the oxide layer formed on the metal alloy, ε_0 is the vacuum permittivity, which has a value of 8.85×10^{-14} F·cm^{-1}, and g is a numerical coefficient close to 1 when α is 1, which can be estimated using the following equation:

$$g = 1 + 2.88\,(1 - a)^{2.375} \quad (5)$$

Furthermore, ρ_0 and ρ_δ represent the lower and upper limits where the CPE behavior is observed in the frequency range. It should be noted that the Q value also corresponds to a parameter of the CPE, which relates to the dielectric constant, resistivity, and film thickness. The following equation can determine it:

$$Q = \frac{(\varepsilon\epsilon_0)^a}{g\delta\rho_{\delta}^{1-a}} \quad (6)$$

The graphical method for analyzing impedance data allowed us to estimate the Constant Phase Element (CPE) parameters and electrolytic resistance (R_e). For the 0% CeO_2-NPs sample, the Q coefficient was determined to be 6.87×10^{-7} F·cm^{-2}·s$^{-(1-\alpha)}$, with $|\alpha|$ corresponding to 0.44. In contrast, for 1% CeO_2-NPs sample, the Q coefficient was found to be 3.62×10^{-7} F·cm^{-2}·s$^{-(1-\alpha)}$ with an $|\alpha|$ value of 0.50. The electrolytic resistance (R_e) was assessed using a graphical method, resulting in approximately 0.935 Ω cm^2 values for the 0% CeO_2-NPs sample and 2.73 Ω cm^2 for the 1% CeO_2-NPs sample. These results demonstrate that adding nanoparticles results in an anodic behavior, accompanied by a re-passivation process at extended exposure times.

Figure 9a shows an equivalent circuit proposed for 0% CeO_2-NPs and 1% CeO_2-NPs after exposure to Li_2CO_3-K_2CO_3 at 550 °C under an H_2/N_2 atmosphere. Additionally, Figure 9b) shows the fitting results obtained by adjusting the experimental data, with χ^2 = 1.8. It should be noted that the accuracy of the fitting is comparable to that of the graphic method.

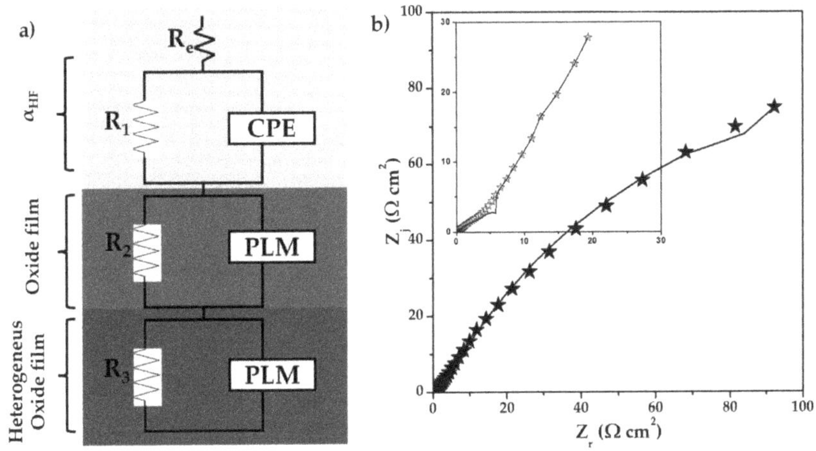

Figure 9. (**a**) A representative equivalent circuit of Cu-50Ni-5Al, (**b**) a comparison of ((★) 0% CeO_2–NPs, (☆) 1% CeO_2–NPs) experimental, and (—) fitting adjusting of impedance data after 120 h of exposure in Li2CO3-K_2CO_3 at 550 °C under an H_2/N_2 atmosphere.

3.3. Morphological and Microstructural Analysis

3.3.1. Scanning Electron Microscopy Analysis

The microstructure of the samples exposed to molten carbonate (Li_2CO_3-K_2CO_3 62–38 mol.%) in an H_2/N_2 atmosphere is presented in Figure 10. Both alloys exhibit a mixed microstructure characterized by filament-shaped particles and a heterogeneous base corresponding to the surface's aluminum and copper oxides. A significant surface modification was observed in the samples containing CeO_2-NPs, leading to an increased formation of cracks and pores and a heterogeneous surface associated with the deposition of molten carbonates in the corrosion products. Additionally, the 1% CeO_2-NPs sample exhibits a low presence of pores in the perforated area of the sample (see Figure 10f). Bright areas resembling needle-like structures corresponding to the microstructure of the molten carbonates were also observed.

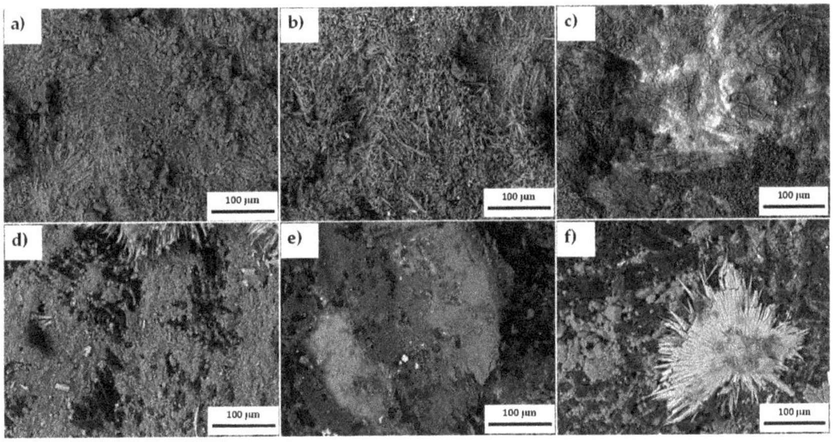

Figure 10. SEM images of the alloys (**a**–**c**) 0% CeO_2-NPs, and (**d**–**f**) 1% CeO_2-NPs.

SEM/EDS mapping, as shown in Figure 11, provided information on the elemental distribution within both alloys, revealing areas with different shades. After 120 h of exposure, molten carbonates prevail on the surface of both alloys, concentrating oxygen and potassium in their vicinity (Figure 11a,b). Based on these results, the surface of the test pieces was cleaned to remove carbonates and to observe the heterogeneity of the elements that compose the alloy. Figure 11c,d show a clear segregation of the different elements resulting from forming their corrosion products (presented in XRD; see Figure 12).

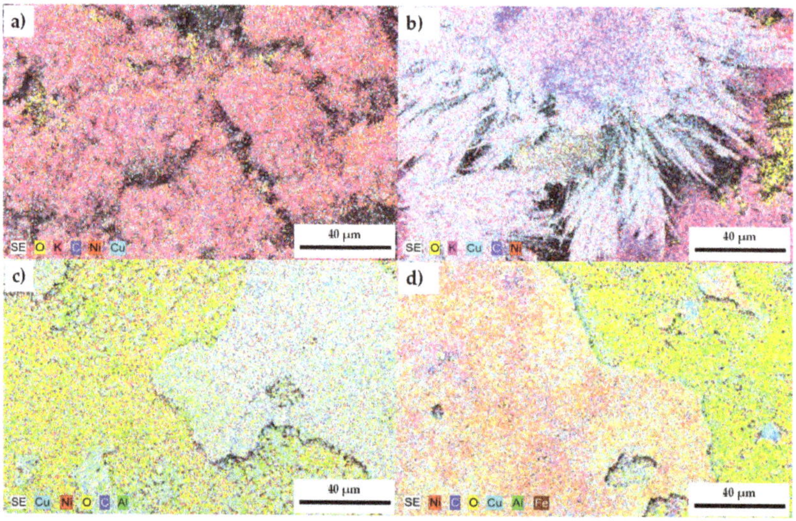

Figure 11. Mapping images of 0% CeO_2-NPs and 1% CeO_2-NPs. (**a**,**b**) After exposure to Li_2CO_3-K_2CO_3 at 550 °C under an H_2 atmosphere and (**c**,**d**) with surface washing.

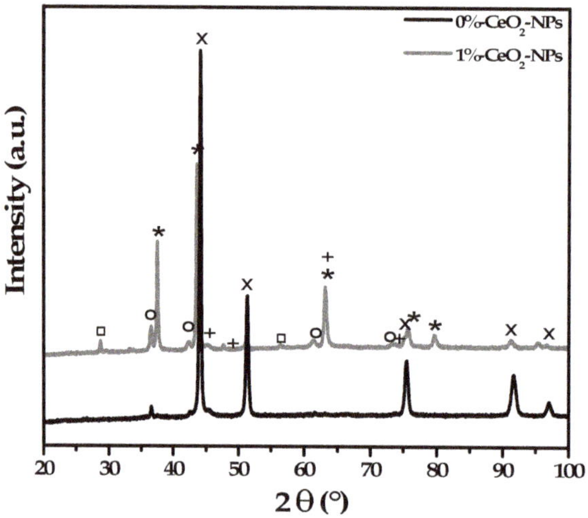

Figure 12. XRD patterns of 0% CeO_2-NPs and 1% CeO_2–NPs, after exposure to Li_2CO_3–K_2CO_3 at 550 °C under an H_2/N_2 atmosphere. (x) CuNiAl-SS; (□) CeO_2; (*) NiO; (o) Cu_2O; (+) Al_2O_3.

At the grain boundaries, a higher concentration of copper is observed (Figure 11d), probably due to the natural tendency of copper–nickel alloys to undergo phase separation, indicated by the miscibility gap in their binary phase diagram [58]. This is accentuated in the samples by incorporating CeO_2-NPs, where a Ni-rich zone, rich in Al and Cu, is observed. Furthermore, the formation of copper and nickel oxides in these regions may also contribute to the observed differences. This agrees with the results of EIS.

3.3.2. X-Ray Diffraction Analysis

Figure 12 presents the diffraction patterns of the 0% CeO_2-NPs and 1% CeO_2-NPs after 120 h of exposure to molten carbonate. In the 0% CeO_2-NPs sample, the diffraction pattern reveals a Cu-Ni-Al solid solution identified as typical fcc structures (Fm-3m) and minor phases of Cu_2O (Pn-3m; JCPDS 010751531), and Al_2O_3 (R-3c; JCPDS 010772135). However, in the 1% CeO_2-NPs NPs sample, the pattern shows the Cu-Ni-Al solid solution, the intriguing addition of CeO_2 (Fm-3m; JCPDS 010750076), and the Cu_2O and Al_2O_3 phases. Notably, NiO (Fm-3m; JCPDS 010731519) is also present, a surprising element absent in the absence of NPs. This is likely due to improved electronic coupling and interatomic interactions between CeO_2 and NiO, which promote the kinetics of the process and electrochemical degradation [59].

3.3.3. X-Ray Photoelectron Spectroscopy Analysis

Figure 13 presents the complete XPS spectrum of the 0% CeO_2-NPs and 1% CeO_2-NPs. In the case of the 0% CeO_2-NPs sample, the deconvolution of the peaks reveals signals related to $Cu(OH)_2$ in the Cu 2p spectrum at binding energies of 935.9 eV, 940.9 eV, 942.3 eV, 943.8 eV, and 944.9 eV. $Cu(OH)_2$ possesses numerous electrochemically active sites, excellent energy storage capabilities, and a high surface area. Recent studies have shown its potential applications as a bacterial photocatalyst and in the electrochemical storage of hydrogen [60]. The Al 2p signal is observed as Al metal (at binding energies of 72 eV and 74.2 eV) and overlaps with Al_2O_3 (75.89 eV), indicating the presence of a passive layer on the surface of the alloy [61]. Notably, the Ni 2p signal was not detected in the XPS analysis. In corrosive environments, copper oxides (CuO, Cu_2O) form preferentially on the surface of Cu-Ni alloys, while nickel can be incorporated into the copper oxide structure or remain on the substrate, making it difficult to detect by XPS [62]. Therefore, the primary corrosion products in this alloy consist of copper oxides (both I and II) and aluminum oxide.

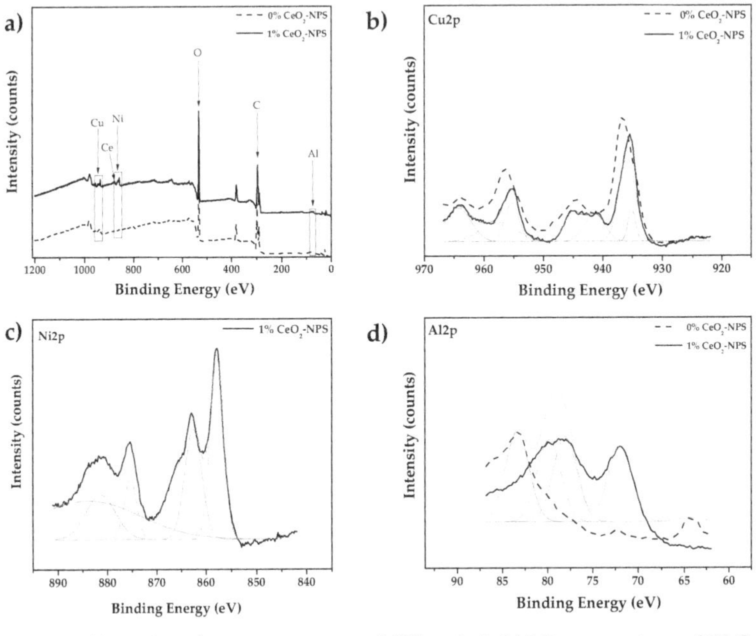

Figure 13. X-ray photoelectron spectroscopy (XPS) analysis (**a**) full scan spectrum of 0% CeO_2-NPs and 1% CeO_2–NPs, (**b**) Cu 2p, (**c**) Ni 2p, and (**d**) Al 2p.

The deconvolution of the peaks for the 1% CeO_2-NPs sample shows the presence of CuO at binding energies of 940.9 eV, 942.4 eV, and 944 eV, as well as Cu_2O at 935.4 eV and 956.0 eV. This suggests that a passive layer is likely to form on the surface. The Al 2p signal corresponds to Al_2O_3, exhibiting peaks at 75.0 eV and 89.0 eV with low intensity, likely due to the small amount of aluminum present in the alloy. The Ni 2p signal is identified as nickel hydroxide ($Ni(OH)_2$) in the presence of CeO_2-NPs with binding energies of 857.8 eV, 863.0 eV, and 864.7 eV, and the twin peaks of Ce^{3+} and Ce^{4+} can be observed at 882.6 eV and 885.5 eV, respectively [63], as shown in Figure 13c. The variable oxidation states of cerium (Ce) contribute to an increase in oxygen vacancies. The presence of Ce^{3+} ions facilitates interactions between ceria and surrounding nickel, thanks to the lone electrons in the 4f orbital. Oxygen-rich cerium oxide can also donate lattice oxygen to nickel, helping maintain the nickel ion in a higher valence of Ni^{2+} in $NiCeO_x$. This process enhances the catalytic activity of the proposed material as anode [64–66]. According to Wang et al. [64], the presence of oxygen vacancy clusters in CeO_2 nanocubes significantly enhances its catalytic activity, which is attributed to improved oxygen mobility and surface reactivity. These oxygen vacancies serve as active sites that facilitate the adsorption and activation of oxygen species, which in turn promote oxidative reactions on the catalyst surface. Moreover, Kumar et al. [65] demonstrated that the concentration of Ce^{3+} states and surface oxygen vacancies strongly influences the electrochemical performance of CeO_2 nanoparticles. Their work emphasizes that Ce^{3+} not only enhances the electrical conductivity of the material but also increases the reactivity of oxygen species at the surface. These activated oxygen species are crucial for the oxidation of Ni species, facilitating the stabilization and formation of NiO. In alignment with these findings, Xiao et al. [66] explored the effect of Pr doping in CeO_2 to engineer oxygen vacancies and improve the dispersion of nickel species. They observed that the presence of Pr further increased the density of oxygen vacancies, enhancing the stability and catalytic efficiency of the CeO_2-NiO composite. This supports the hypothesis that CeO_2 actively contributes to the formation and stabilization of NiO through its redox properties and oxygen mobility. These findings align with several studies indicating that NiO, NiO(OH), and $Ni(OH)_2$ are suitable materials for use as electrodes in fuel cells and batteries due to their catalytic properties [67–69], which confirms the results observed in the SEM images.

4. Conclusions

Mechanical alloying and spark plasma sintering (SPS) were employed to produce Cu–50Ni–5Al alloys reinforced with and without 1% CeO_2 nanoparticles (NPs).

The progressive loading scratch tests demonstrate that adding 1 wt.% nanoparticles can significantly reduce the coefficient of friction by up to 50%. Furthermore, the presence of nanoparticles leads to shallower scratches, indicating an improvement in wear resistance. The findings further indicate that key material characteristics—such as porosity, hardness, and nanoparticle incorporation—significantly shape the material's response to single scratch events.

Electrochemical impedance spectroscopy (EIS) assessed their electrochemical behavior on a molten carbonate (Li_2CO_3-K_2CO_3) in a hydrogen atmosphere at 550 °C for 120 h. The results indicated that incorporating CeO_2 nanoparticles reduces polarization resistance but promotes the re-passivation of the material at a longer exposure time.

The formed corrosion products were analyzed using SEM-EDS, X-ray diffraction and X-ray photoelectron spectroscopy. In both systems, the corrosion products consisted of aluminum oxide and copper oxide. Furthermore, incorporating nanoparticles in an H_2/N_2 atmosphere decisively promotes the formation of nickel oxides as the predominant phase.

Therefore, incorporating CeO_2 nanoparticles into the Cu-50Ni-5Al system increased wear resistance. Additionally, it produces an anodic behavior and a re-passivation process in a hydrogen atmosphere at long exposure times.

Understanding this impact is crucial for developing strategies to enhance the material's durability in challenging environments such as MCFC.

Author Contributions: Conceptualization, C.M. and B.V.; methodology, C.M. and F.B.; software, C.M. and B.V.; validation, C.M.; formal analysis, C.M. and B.V.; investigation, C.M., B.V., F.B., A.D.V.-R. and B.B.-D.L.F.; resources, C.M.; data curation, B.V., F.B., A.D.V.-R. and B.B.-D.L.F.; writing—original draft preparation, C.M., B.V. and F.B.; writing—review and editing, C.M., B.V. and I.F.M.; visualization, C.M., B.V. and I.F.M.; supervision, C.M.; project administration, C.M.; funding acquisition, C.M. All authors have read and agreed to the published version of the manuscript.

Funding: This research received no external funding.

Institutional Review Board Statement: Not applicable.

Informed Consent Statement: Not applicable.

Data Availability Statement: The data presented in this study are available on request from the corresponding author.

Acknowledgments: The authors thank ANID FONDECYT Grant 11190500 and FONDEQUIP EQM130014. EQM 170087. Millennium Institute on Green Ammonia as Energy Vector—MIGA (ICN2021_023) supported by the Millennium Scientific Initiative by ANID. The authors thank Brazilian foundations (CNPq, CAPES, and FAPESP).

Conflicts of Interest: The authors declare no conflicts of interest.

Abbreviations

The following abbreviations are used in this manuscript:

SPS	Spark plasma sintering
EIS	Electrochemical impedance spectroscopy
NP	Nanoparticles
SEM	Scanning electron microscopy
XPS	X-ray photoelectron spectroscopy
XRD	X-ray diffraction
MCFCs	Molten carbonate fuel cells
SOFCs	Solid oxide fuel cells
Cu	Copper
Al	Aluminum
Ni	Nickel
CeO_2	Cerium oxide
ICR	Interface contact resistance
BPR	Ball-to-powder ratio
F_t	Tangential force
F_n	Normal force
COF	Coefficient of friction
E_{OC}	Open circuit potential
HF	High frequency
LF	Low frequency
C_{dl}	Capacitance of the electric double-layer
CPE	Constant phase element
PLM	Power low model
R_e	Electrolytic resistance

References

1. Manzo, D.; Thai, R.; Le, H.T.; Venayagamoorthy, G.K. Fuel cell technology review: Types, economy, applications, and vehicle-to-grid scheme. *Sustain. Energy Technol. Assess.* **2025**, *75*, 104229. [CrossRef]
2. Wejrzanowski, T.; Cwieka, K.; Skibinski, J.; Lysik, A.; Ibrahim, S.H.; Milewski, J.; Xing, W.; Lee, C.-G. Microstructure driven design of porous electrodes for molten carbonate fuel cell application: Recent progress. *Int. J. Hydrogen Energy* **2020**, *45*, 25719–25732. [CrossRef]
3. Ang, S.M.C.; Fraga, E.S.; Brandon, N.P.; Samsatli, N.J.; Brett, D.J. Fuel cell systems optimisation—Methods and strategies. *Int. J. Hydrogen Energy* **2011**, *36*, 14678–14703. [CrossRef]
4. Cigolotti, V.; Genovese, M.; Fragiacomo, P. Comprehensive Review on Fuel Cell Technology for Stationary Applications as Sustainable and Efficient Poly-Generation Energy Systems. *Energies* **2021**, *14*, 4963. [CrossRef]
5. Qasem, N.A.A.; Abdulrahman, G.A.Q. A Recent Comprehensive Review of Fuel Cells: History, Types, and Applications. *Int. J. Energy Res.* **2024**, *2024*, 7271748. [CrossRef]
6. Rana, M.; Uddin, M.; Sarkar, R.; Meraj, S.T.; Shafiullah, G.; Muyeen, S.; Islam, A.; Jamal, T. Applications of energy storage systems in power grids with and without renewable energy integration—A comprehensive review. *J. Energy Storage* **2023**, *68*, 107811. [CrossRef]
7. Tomczyk, P. MCFC versus other fuel cells—Characteristics, technologies and prospects. *J. Power Sources* **2006**, *160*, 858–862. [CrossRef]
8. Mehr, A.; Lanzini, A.; Santarelli, M.; Rosen, M.A. Polygeneration systems based on high temperature fuel cell (MCFC and SOFC) technology: System design, fuel types, modeling and analysis approaches. *Energy* **2021**, *228*, 120613. [CrossRef]
9. Jang, S.-C.; Lee, B.Y.; Nam, S.W.; Ham, H.C.; Han, J.; Yoon, S.P.; Oh, S.-G. New method for low temperature fabrication of Ni–Al alloy powder for molten carbonate fuel cell applications. *Int. J. Hydrogen Energy* **2014**, *39*, 12259–12265. [CrossRef]
10. Li, G.; Thomas, B.G.; Stubbins, J.F. Modeling creep and fatigue of copper alloys. *Metall. Mater. Trans. A Phys. Metall. Mater. Sci.* **2000**, *31*, 2491–2502. [CrossRef]
11. Klassert, A.; Tikana, L. Copper and copper-nickel alloys—An overview. In *Corrosion Behaviour and Protection of Copper and Aluminium Alloys in Seawater*; CRC Press: Boca Raton, FL, USA, 2007; pp. 47–61. [CrossRef]
12. Abdollahipour, A.; Sayyaadi, H. Thermal energy recovery of molten carbonate fuel cells by thermally regenerative electrochemical cycles. *Energy* **2021**, *227*, 120489. [CrossRef]
13. Appleby, A.J.; Blomen, L.J.M.J.; Mugerwa, M.N. (Eds.) *Fuel Cell Systems*; Plenum Press: New York, NY, USA, 1993; pp. 157–197.
14. Czelej, K.; Cwieka, K.; Colmenares, J.C.; Kurzydlowski, K.J. Catalytic activity of NiO cathode in molten carbonate fuel cells. *Appl. Catal. B Environ.* **2018**, *222*, 73–75. [CrossRef]
15. Wee, J.-H.; Song, D.-J.; Jun, C.-S.; Lim, T.-H.; Hong, S.-A.; Lim, H.-C.; Lee, K.-Y. Evaluation of Ni-Ni_3Al(5 wt.%)-Al(3 wt.%) as an anode electrode for molten carbonate fuel cell: Part I: Creep and sintering resistance. *J. Alloys Compd.* **2005**, *390*, 155–160. [CrossRef]
16. Hwang, E.; Park, J.; Kim, Y.; Kim, S.; Kang, S. Effect of alloying elements on the copper-base anode for molten carbonate fuel cells. *J. Power Sources* **1997**, *69*, 55–60. [CrossRef]
17. Li, C.; Chen, J.; Li, W.; Hu, Y.; Ren, Y.; Qiu, W.; He, J.; Chen, J. Investigation on compressive behavior of Cu-35Ni-15Al alloy at high temperatures. *Mater. Sci.* **2014**, *32*, 341–349. [CrossRef]
18. Martínez, C.; Arcos, C.; Briones, F.; Machado, I.; Sancy, M.; Bustamante, M. The Effect of Adding CeO_2 Nanoparticles to Cu-Ni-Al Alloy for High Temperatures Applications. *Nanomaterials* **2024**, *14*, 143. [CrossRef]
19. Hume-Rothery, W. The Structure of Metals and Alloys. *Nature* **1936**, *138*, 7–8.
20. Elorz, J.A.P.-S.; Hernández, M.J.Q.; González, L.F.V.; Hernández, Q.; González, L.F.V. Total Insolubility and Solubility in Alloys. In *Solidification and Solid-State Transformations of Metals and Alloys*; Elsevier: Amsterdam, The Netherlands, 2017; pp. 67–92. [CrossRef]
21. Accardo, G.; Frattini, D.; Moreno, A.; Yoon, S.P.; Han, J.H.; Nam, S.W. Influence of nano zirconia on NiAl anodes for molten carbonate fuel cell: Characterization, cell tests and post-analysis. *J. Power Sources* **2017**, *338*, 74–81. [CrossRef]
22. Accardo, G.; Frattini, D.; Yoon, S.P.; Ham, H.C.; Nam, S.W. Performance and properties of anodes reinforced with metal oxide nanoparticles for molten carbonate fuel cells. *J. Power Sources* **2017**, *370*, 52–60. [CrossRef]
23. Jomjaree, T.; Sintuya, P.; Srifa, A.; Koo-Amornpattana, W.; Kiatphuengporn, S.; Assabumrungrat, S.; Sudoh, M.; Watanabe, R.; Fukuhara, C.; Ratchahat, S. Catalytic performance of Ni catalysts supported on CeO_2 with different morphologies for low-temperature CO_2 methanation. *Catal. Today* **2021**, *375*, 234–244. [CrossRef]
24. Mustafa, S.Z.; Qureshi, M.D.; Jabeen, I.; Khalid, S.; Hussain, G.; Riaz, J.; Akram, M.N.; Maqsood, F. Physiochemical Role of Nanoparticles in Solid Fuel Cells, Production and Applications in Physics and Chemical Sciences. *Sch. Bull.* **2022**, *8*, 66–70. [CrossRef]
25. Gong, B.; Ma, X.; Wang, T.; Hou, J.; Ji, S.; Xu, Q.; Cao, H. Emerging two dimensional MXene for corrosion protection in new energy systems: Design and mechanisms. *Adv. Colloid Interface Sci.* **2024**, *336*, 103373. [CrossRef] [PubMed]

26. Jang, G.G.; Jun, J.; Keum, J.K.; Su, Y.-F.; Pole, M.; Niverty, S.; Joshi, V.V. Lithium nitrate salt-assisted CO_2 absorption for the formation of corrosion barrier layer on AZ91D magnesium alloy. *RSC Adv.* **2024**, *14*, 17696–17709. [CrossRef] [PubMed]
27. Venkatesh, G.; Gnanamoorthy, R.; Okazaki, M. Fretting wear behaviour of nickel foam struts used in fuel cell applications. Proceedings of the Institution of Mechanical Engineers. *Part J J. Eng. Tribol.* **2021**, *236*, 144–155. [CrossRef]
28. Pukha, V.E.; Glukhov, A.; Belmesov, A.A.; Kabachkov, E.N.; Khodos, I.I.; Khadem, M.; Kim, D.E.; Karaseov, P.A. Corrosion-resistant nanostructured carbon-based coatings for applications in fuel cells based on bipolar plates. *Vacuum* **2023**, *218*, 112643. [CrossRef]
29. Xu, N.; Ma, J.; Liu, Q.; Han, W.; Shan, Z. Size Effect of CeO_2 Particle on Nanoscale Single-Asperity Sliding Friction. *Tribol. Lett.* **2022**, *70*, 4. [CrossRef]
30. Mishra, T.K.; Kumar, A.; Sinha, S.K.; Gupta, B. Wear behavior and XRD analysis of reinforced copper matrix composite re-inforced with Cerium Oxide (CeO_2). *Mater. Today Proc.* **2018**, *5*, 27786–27794. [CrossRef]
31. Xu, N.; Han, W.; Wang, Y.; Li, J.; Shan, Z. Nanoscratching of copper surface by CeO_2. *Acta Mater.* **2017**, *124*, 343–350. [CrossRef]
32. Yu, Z.; Chen, M.; Wang, Q.; Wang, F. A CeO_2-Doped Copper Matrix Self-Lubricating Composite with Well-Balanced Mechanical and Lubricating Properties Fabricated by Spark Plasma Sintering. *Met. Mater. Trans. A* **2022**, *53*, 2665–2678. [CrossRef]
33. Martínez, C.; Briones, F.; Araya, N.; Aguilar, C.; Machado, I.; Guerra, C.; Medina, A.; Sancy, M. Influence of the synthesis technique on tribological behavior of a Ti-6Al-4V alloy. *Mater. Lett.* **2020**, *281*, 128627. [CrossRef]
34. Bergami, L.B.; Lima, A.O.; Venturelli, B.N.; Machado, I.F.; Albertin, E.; Souza, R.M. Effect of carbide orientation during single scratch test in directionally solidified and heat-treated high chromium cast irons. *Wear* **2023**, *523*, 204823. [CrossRef]
35. Mangam, V.; Bhattacharya, S.; Das, K.; Das, S. Friction and wear behavior of Cu-CeO_2 nanocomposite coatings synthesized by pulsed electrodeposition. *Surf. Coat. Technol.* **2010**, *205*, 801–805. [CrossRef]
36. Shi, Z.; Meng, Y. Effects of indentation depth and grain size on scratching behavior of nanograin FCC Fe polycrystalline substrate. *Tribol. Int.* **2024**, *193*, 109464. [CrossRef]
37. Holmberg, K.; Matthews, A.; Ronkainen, H. Coatings tribology: Properties, techniques and applications in surface engineering. *Tribol. Int.* **1998**, *31*, 107–120. [CrossRef]
38. Roy, M.; Singh, S.K.; Das, S. Influence of hardness and toughness on wear resistance of materials. *Tribol. Int.* **2018**, *119*, 10–20.
39. Zum Gahr, K.H. *Microstructure and Wear of Materials*; Elsevier: Amsterdam, The Netherlands, 1987.
40. Chen, J.; Shen, Y.; Liu, W.; Beake, B.D.; Shi, X.; Wang, Z.; Zhang, Y.; Guo, X. Effects of loading rate on development of pile-up during indentation creep of polycrystalline copper. *Mater. Sci. Eng. A* **2016**, *656*, 216–221. [CrossRef]
41. Setti, D.; Kirsch, B.; Aurich, J.C. Experimental investigations and kinematic simulation of single grit scratched surfaces considering pile-up behaviour: Grinding perspective. *Int. J. Adv. Manuf. Technol.* **2019**, *103*, 471–485. [CrossRef]
42. Franco, L.A.; Sinatora, A. Material removal factor (f ab): A critical assessment of its role in theoretical and practical approaches to abrasive wear of ductile materials. *Wear* **2017**, *382–383*, 51–61. [CrossRef]
43. Hokkirigawa, K.; Kato, K. An experimental and theoretical investigation of ploughing, cutting and wedge formation during abrasive wear. *Tribol. Int.* **1988**, *21*, 51–57. [CrossRef]
44. Boidi, G.; Fukumasu, N.K.; Machado, I.F. Wear and friction performance under lubricated reciprocating tests of steel powder mixtures sintered by Spark Plasma Sintering. *Tribol. Int.* **2018**, *121*, 139–147. [CrossRef]
45. Fukumasu, N.K.; Boidi, G.; Seriacopi, V.; Machado, G.A.A.; Souza, R.M.; Machado, I.F. Numerical analyses of stress induced damage during a reciprocating lubricated test of fecmo SPS sintered alloy. *Tribol. Int.* **2017**, *113*, 443–447. [CrossRef]
46. Meléndez-Ceballos, A.; Albin, V.; Fernández-Valverde, S.M.; Ringuedé, A.; Cassir, M. Electrochemical Properties of Atomic Layer Deposition Processed CeO_2 as a Protective Layer for the Molten Carbonate Fuel Cell Cathode. *Electrochim. Acta* **2014**, *140*, 174–181. [CrossRef]
47. Fang, Y.; Shu, X.; Dong, S. High Temperature Oxidation Behavior of Nano-Alumina–Modified NiAl Coating. *Front. Mater.* **2022**, *9*, 934215. [CrossRef]
48. Appleby, A.J.; Nicholson, S.B. Reduction of oxygen in lithium-potassium carbonate melt. *J. Electroanal. Chem. Interfacial Electrochem.* **1980**, *112*, 71–76. [CrossRef]
49. Nishina, T.; Masuda, Y.; Uchida, I.; Saboungi, M.-L.; Kojima, H. *Química y Tecnología de Sales Fundidas*; Elec-trochemical Society Proceeding Series; Saboungi, M.-L., Kojima, H., Eds.; Electrochemical Society: Pennington, NJ, USA, 1993; p. 424.
50. Valverde, B.; Sancy, M.; Martínez, C.; Botello, E.; Pío, E.; Aguilar, C. Effect of Sn Content on the Passivity of Ti-Ta-Sn Alloys. *Corros. Sci.* **2023**, *211*, 110887. [CrossRef]
51. Pineda, F.; Walczak, M.; Vilchez, F.; Guerra, C.; Escobar, R.; Sancy, M. Evolution of Corrosion Products on ASTM A36 and AISI 304L Steels Formed in Exposure to Molten $NaNO_3$-KNO_3 Eutectic Salt: Electrochemical Study. *Corros. Sci.* **2022**, *196*, 110047. [CrossRef]
52. Martínez, C.; Guerra, C.; Silva, D.; Cubillos, M.; Briones, F.; Muñoz, L.; Páez, M.A.; Aguilar, C.; Sancy, M. Effect of Porosity on Mechanical and Electrochemical Properties of Ti-6Al-4V Alloy. *Electrochim. Acta* **2020**, *338*, 135858. [CrossRef]

53. Lee, C.-G. Influence of temperature on the anode reaction in a molten carbonate fuel cell. *J. Electroanal. Chem.* **2017**, *785*, 152–158. [CrossRef]
54. Lee, C.-G. Analysis of impedance in a molten carbonate fuel cell. *J. Electroanal. Chem.* **2016**, *776*, 162–169. [CrossRef]
55. Orazem, M.E.; Tribollet, B. Constant-Phase Elements. In *Electrochemical Impedance Spectroscopy*; Wiley: Hoboken, NJ, USA, 2017; pp. 395–419. ISBN 9781118527399.
56. Tribollet, B.; Vivier, V.; Orazem, M.E. *EIS Technique in Passivity Studies: Determination of the Dielectric Properties of Passive Films*; Wandelt, K., Ed.; Elsevier: Amsterdam, The Netherlands, 2018; ISBN 9780128098943.
57. Hirschorn, B.; Orazem, M.E.; Tribollet, B.; Vivier, V.; Frateur, I.; Musiani, M. Constant-Phase-Element Behavior Caused by Resistivity Distributions in Films: II. Applications. *J. Electrochem. Soc.* **2010**, *157*, C458. [CrossRef]
58. Turchanin, M.A.; Agraval, P.G.; Abdulov, A.R. Phase equilibria and thermodynamics of binary copper systems with 3d-metals. VI. Copper-nickel system. *Powder Metall. Met. Ceram.* **2007**, *46*, 467–477. [CrossRef]
59. Patel, K.B.; Mariyaselvakumar, M.; Vyas, G.; Chaudhari, J.C.; Patidar, R.; Srinivasan, K.; Srivastava, D.N.; Bhadu, G.R. Nickel oxide doped ceria nanoparticles (NiO@CeO_2) for boosting oxygen evolution reaction and enhancing stability. *Appl. Surf. Sci.* **2024**, *649*, 159212. [CrossRef]
60. He, D.; Wang, G.; Liu, G.; Bai, J.; Suo, H.; Zhao, C. Facile route to achieve mesoporous $Cu(OH)_2$ nanorods on copper foam for high-performance supercapacitor electrode. *J. Alloys Compd.* **2017**, *699*, 706–712. [CrossRef]
61. Tang, J.; Han, Z.; Zuo, Y.; Tang, Y. A corrosion resistant cerium oxide based coating on aluminum alloy 2024 prepared by brush plating. *Appl. Surf. Sci.* **2010**, *257*, 2806–2812. [CrossRef]
62. Jin, T.; Zhang, W.; Li, N.; Liu, X.; Han, L.; Dai, W. Surface Characterization and Corrosion Behavior of 90/10 Copper-Nickel Alloy in Marine Environment. *Materials* **2019**, *12*, 1869. [CrossRef]
63. Reddy, B.M.; Khan, A.; Yamada, Y.; Kobayashi, T.; Loridant, S.; Volta, J.-C. Surface Characterization of CeO_2/SiO_2 and $V_2O_5/CeO_2/SiO_2$ Catalysts by Raman, XPS, and Other Techniques. *J. Phys. Chem. B* **2003**, *106*, 1520–6106. [CrossRef]
64. Wang, L.; Yu, Y.; He, H.; Zhang, Y.; Qin, X.; Wang, B. Oxygen vacancy clusters essential for the catalytic activity of CeO_2 nanocubes for o-xylene oxidation. *Sci. Rep.* **2017**, *7*, 12845. [CrossRef]
65. Kumar, M.; Yun, J.H.; Bhatt, V.; Singh, B.; Kim, J.; Kim, J.S.; Kim, B.S.; Lee, C.Y. Role of Ce^{3+} valence state and surface oxygen vacancies on enhanced electrochemical performance of single step solvothermally synthesized CeO_2 nanoparticles. *Electro-Chim. Acta* **2018**, *284*, 709–720. [CrossRef]
66. Xiao, Z.; Li, Y.; Hou, F.; Wu, C.; Pan, L.; Zou, J.; Wang, L.; Zhang, X.; Liu, G.; Li, G. Engineering oxygen vacancies and nickel dispersion on CeO_2 by Pr doping for highly stable ethanol steam reforming. *Appl. Catal. B* **2019**, *258*, 117940. [CrossRef]
67. Brisse, A.-L.; Stevens, P.; Toussaint, G.; Crosnier, O.; Brousse, T. $Ni(OH)_2$ and NiO Based Composites: Battery Type Electrode Materials for Hybrid Supercapacitor Devices. *Materials* **2018**, *11*, 1178. [CrossRef]
68. Freitas, M.B.J.G. Nickel Hydroxide Powder for $NiO{\cdot}OH/Ni(OH)_2$ Electrodes of the Alkaline Batteries. *J. Power Sources* **2001**, *93*, 163–173. [CrossRef]
69. Cheng, T.-L.; Lei, Y.; Chen, Y.; Fan, Y.; Abernathy, H.; Song, X.; Wen, Y.-H. Oxidation of Nickel in Solid Oxide Cells during Electrochemical Operation: Experimental Evidence, Theoretical Analysis, and an Alternative Hypothesis on the Nickel Migration. *J. Power Sources* **2023**, *569*, 232991. [CrossRef]

Disclaimer/Publisher's Note: The statements, opinions and data contained in all publications are solely those of the individual author(s) and contributor(s) and not of MDPI and/or the editor(s). MDPI and/or the editor(s) disclaim responsibility for any injury to people or property resulting from any ideas, methods, instructions or products referred to in the content.

materials

MDPI

Article

Mg–Zn–Ca Alloy with Ultra-High Ductility and Strength Processed by Screw Rolling

Haoran Zheng [1,2], Weitao Sun [1], Lijun Deng [1,3], Li Zhao [1], Kwang Seon Shin [4,*] and Jian Zhang [1,*]

1 Aerospace and Mechanical College, Shandong University of Aeronautics, Binzhou 256600, China; zhenghaoran916@163.com (H.Z.); sun_wei_tao@163.com (W.S.); denglijun_piston@163.com (L.D.); zhaoliafr@163.com (L.Z.)

2 Shandong Provincial Key Laboratory of Advanced Technology and Equipment for Laser Additive Manufacturing, Binzhou 256600, China

3 Shandong Key Laboratory of Advanced Engine Piston Assembly, Binzhou 256600, China

4 Magnestum Technology Innovation Center, School of Materials Science and Engineering, Seoul National University, Seoul 08826, Republic of Korea

* Correspondence: ksshin@snu.ac.kr (K.S.S.); zhangjian3829@163.com (J.Z.)

Abstract: Mg alloys are highly attractive for biodegradable surgical clips because of their low density and good biocompatibility; however, their limited strength and ductility restrict their widespread application. To overcome this limitation, this study employed screw rolling (SR) to produce a Mg–3Zn–0.2Ca alloy with a fine microstructure and an average grain size of 1.6 μm. Experimental results showed that the SR process improved the comprehensive tensile properties of the alloy, increasing the yield strength, ultimate tensile strength, and elongation from 192.6, 234.4 MPa, and 21.7% for the pre-extruded alloy to 252.3, 289 MPa, and 39.5%, respectively. Quantitative analysis of the strengthening behaviour identified grain refinement as the primary strengthening mechanism, along with considerable contributions from Orowan and dislocation strengthening. The ultra-high-tensile ductility was primarily attributed to the low internal stress, nano-sized precipitates, texture weakening, and activation of multiple slip systems. These findings provide a strategy for simultaneously increasing the ductility and strength of Mg alloys and lay a foundation for applying them as biodegradable clips.

Keywords: Mg–Zn–Ca alloy; screw rolling; microstructure; precipitation; mechanical properties

1. Introduction

Biodegradable clips for laparoscopic surgery have attracted widespread attention because they can avoid additional pain and injury caused by removal. Among the different types of surgical clips, the mechanical strength of biodegradable polymer clips is relatively low, and their structure is relatively complex, which limits their application [1,2]. In recent years, Mg and its alloys have garnered growing attention in the realm of biomedical materials because of their excellent biocompatibility and biodegradability [3–6]. However, pure Mg has limited clinical viability because of its low ductility and strength, which are attributed to its hexagonal close-packed (HCP) structure [7]. Adding alloying elements or imposing severe plastic deformation are both effective methods for improving these mechanical properties [8].

Research has shown that Mg–Zn–Ca alloys are suitable for such biodegradable clips [9]. Zn and Ca are important minerals for the human body, offering superior compatibility with human tissues compared to Al and rare-earth elements [10,11]. Incorporating Zn and Ca

 https://doi.org/10.3390/ma18112586

can strengthen Mg alloys through both grain refinement and solid-solution mechanisms; additionally, they form uniformly distributed secondary-phase distributions, which further boost the alloy's mechanical properties [12]. Moreover, when Zn and Ca are incorporated into Mg alloys, together with the internal stress concentration caused by extrusion, they activate the <c + a> slip, thereby enhancing the ductility of the resultant Mg–Zn–Ca alloy [13].

Meanwhile, some researchers have plastically deformed Mg–Zn–Ca alloys to improve their mechanical properties. For example, a Mg–1.0Zn–0.5Ca alloy extruded at 370 °C demonstrated a high elongation of 44%, but its yield strength (YS, 105 MPa) and ultimate tensile strength (UTS, 205 MPa) were low [14]. By contrast, a Mg–0.6Zn–0.5Ca alloy, processed via double equal channel angular pressing (D-ECAP) at 280 °C, exhibited a higher YS of 372 MPa but a lower elongation of 7% [15]. In addition, significant residual stress occurs at the apex of surgical clips during closure, easily leading to cracking [16]. Therefore, optimising the processing method to improve the plasticity and overcome the strength–ductility trade-off of alloys for such clips is important.

Screw rolling (SR) is an emerging deformation processing method with considerable application potential that can refine alloy grains by inducing a large plastic deformation [17–21]. Notably, SR always produces parts with smooth surfaces, which is a considerable advantage of this method. Further, the area reduction achieved by one rolling pass of an SR mill is approximately equal to that obtained by 6–8 rolling passes in a traditional mill, indicating its higher efficiency [22]. The SR deformation process differs from extrusion or rolling but is similar to torsion deformation. As the rolls rotate to deform the billet, it also moves forward in a spiralling motion, which enables the mill to greatly reduce the billet's cross-sectional area. The heat generated by this large plastic deformation and friction may cause dynamic recrystallisation (DRX). Thus, the SR process enables the production of materials with fine-grained structures and enhanced mechanical properties [23].

However, to the best of our knowledge, the SR technique has only rarely been applied to Mg alloys [24–26]. When pure Mg [24] was processed using SR technology, its YS increased from 55 to 146 MPa after 11 SR passes, demonstrating the efficacy of SR for strengthening Mg. In another study, SR processing increased the YS of a pre-extruded Mg-3.71Zn–2.04Al–0.63Ca–0.62Mn alloy [25] from 225.3 to 272.7 MPa because of grain refinement and the dissolution of the secondary phases; however, its elongation decreased from 23.9 to 8.3%. Meanwhile, the fracture elongation of a Mg–5.56%Zn–0.82%Mn alloy [26] increased from 15.2% for the as-received samples to 27.8% after nine SR passes at 300 °C; this increase of 82.9% was primarily attributed to its microstructure and weakened texture. In addition, SR promoted DRX and decreased the elevated dislocation density that was initially induced by work hardening in the as-received sample. Moreover, the YS increased from 115 to 189 MPa through grain refinement and precipitation strengthening. Notably, the final microstructure and texture of a Mg alloy are influenced not just by the alloying elements but also by the processing parameters, such as the number of processing passes and the processing temperature.

Our previous studies showed that the Mg–3Zn–0.2Ca (wt.%) alloy has good biocompatibility, and its YS and elongation were 205 MPa and 17.85%, respectively [27,28]. Therefore, in this study, a Mg–3Zn–0.2Ca alloy was pre-extruded and then subjected to SR to improve its strength and plasticity. An analysis was conducted on the microstructure, texture, secondary-phase, and mechanical performance of the SR-processed Mg–Zn–Ca alloy. The outcomes offer a fresh strategy for boosting the mechanical properties of Mg-based alloys.

2. Experimental Details

This study employed a pure Mg (99.99%) ingot, Zn particles (99.99%, Ke Wei company of Tianjin University Co. Ltd., Tianjin, China), and a Mg–25Ca intermediate alloy as the raw materials. The Mg–3Zn–0.2Ca alloy was synthesized using a vacuum induction melting furnace (Shanghai Chenhua Science Technology Co. Ltd., ZG-10, Shanghai, China) in an Ar atmosphere. The raw materials were melted and stirred at 720 °C for 5 min to homogenise the composition before casting in a graphite mould at 690 °C. After cooling naturally in the furnace, rod billets with a 60 mm diameter were produced. Next, they underwent a homogenization process at 400 °C for 24 h. Following this, a hot extrusion was carried out at 320 °C using a YQ32-315 extruder (Shangdong DaYin Industry Machine Co., Ltd., Jinan, China), with an extrusion ratio of 6:1, resulting in a pre-extruded bar measuring ∅25 mm. The SR process [29] was then performed, with a 30-min anneal before each pass in the sequence of 320, 300, 280, 260, 240, 220, and 200 °C [24]. Finally, a ∅16 mm SR bar was obtained. Figure 1 schematically illustrates the process for preparing the alloy. Hereafter, the specimens identified as 3ZX and SR-3ZX correspond to the pre-extruded and SR Mg–3Zn–0.2Ca (wt%) samples, respectively.

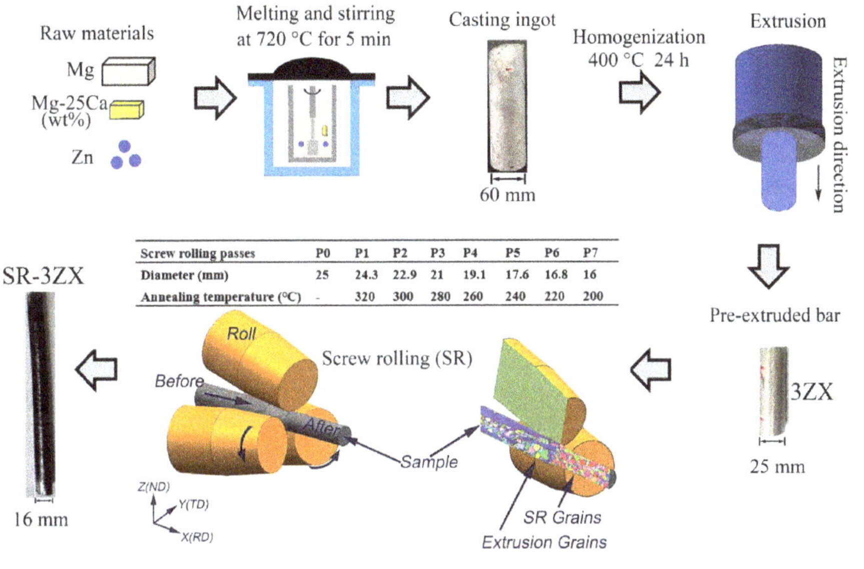

Figure 1. Schematic of the casting, pre-extrusion, and SR processes of the Mg–Zn–Ca alloy.

To observe the microstructures of the alloy specimens, the test specimens were sectioned perpendicular to the direction of extrusion. The microstructure was examined using optical microscopy (OM; OLYMPUS, U-TV0.5XC-3, Tokyo, Japan), scanning electron microscopy (SEM; Helios NanoLab 460HP, Hillsboro, OR, USA), transmission electron microscopy (TEM; Talos F200X, Thermo Fisher Scientific, Waltham, MA, USA), and electron backscatter diffraction (EBSD; Oxford Instruments, NordlysNano, Oxford, UK). For the EBSD analysis, OIM (https://www.edax.com/products/ebsd/oim-analysis, accessed on 8 May 2025) and ATEX software were employed. The grain size was determined by applying the linear intercept method to the obtained EBSD maps.

The hardness values of the specimens were measured using a Vickers hardness tester (HMV-2T, Shimadzu, Kyoto, Japan) with a load of 9.8 N for 20 s. Each specimen was randomly tested in at least six locations, and the values were averaged. Specimens for tensile testing were crafted from a location 0.4 mm from the edge of the bar in the extrusion direction, following the dimensions depicted in Figure 2. The mechanical performance was assessed by tensile tests using a DNS universal testing machine with a strain rate of 10^{-4} s^{-1} at ambient temperature. To ensure accuracy, three identical samples were tested. The fracture surface was observed via SEM, and energy-dispersive X-ray spectroscopy (EDS) was used to analyse the chemical composition of the sample and fracture surfaces. A typical TEM dual-beam analysis was carried out to identify the types of dislocation activation.

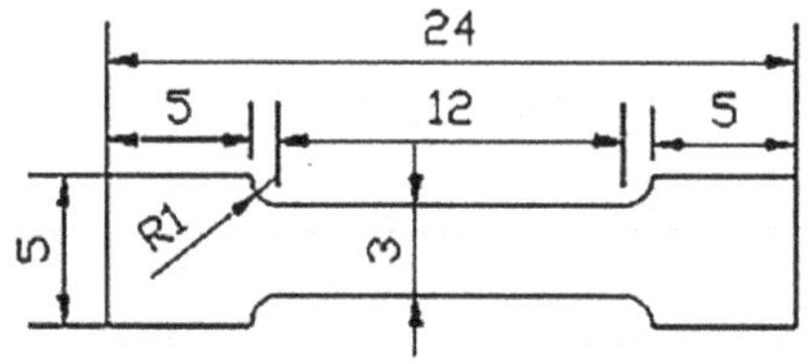

Figure 2. Dimensions of the tensile test samples (mm).

3. Results

3.1. Tensile Properties

Figure 3a illustrates the true stress–strain curves for 3ZX and SR-3ZX, and Figure 3b,c plot their corresponding mechanical performance parameters. The 3ZX specimen exhibits a low UTS of ~234.4 MPa, YS of ~192.6 MPa, and elongation of ~21.7%. Remarkably, for SR-3ZX, these values increase simultaneously to ~289 MPa, ~252.3 MPa, and ~39.5%, respectively. Furthermore, the strain hardening coefficients (n) of 3ZX and SR-3ZX are 0.16 and 0.25, respectively. An increased value of n indicates that the material is capable of withstanding greater strain hardening [30,31]; thus, SR-3ZX also exhibits superior plasticity. In addition, the hardness of 3ZX increased from ~70 to ~82 HV after SR. Overall, SR-3ZX demonstrates favourable mechanical performance, successfully balancing robust strength with remarkable ductility.

Figure 4 shows the fracture surfaces of 3ZX and SR-3ZX following tensile tests conducted at room temperature. As shown in Figure 4a,b, the 3ZX fracture surface exhibits quasi-cleavage fracture characteristics, containing cleavage planes (blue arrows) and a few dimples (yellow arrows). However, as shown in Figure 4c,d, the fracture morphology of SR-3ZX exhibits microvoid coalescence, as evidenced by the presence of many dimples and the absence of typical cleavage planes. Compared with those of SR-3ZX, the 3ZX dimples are smaller and shallower. These phenomena confirm that the SR process significantly increases the ductility of the alloy, as demonstrated by these typical features of a ductile fracture surface. In addition, the green arrow in Figure 4d shows that the precipitates are uniformly distributed at the bottoms of the dimples in the SR-3ZX fracture surface. Further, EDS analysis reveals that the chemical composition of point 1 indicated by the green arrow is 69.87 at.% Mg, 17.62 at.% Zn, and 12.51 at.% Ca, implying that these particles are $Ca_2Mg_6Zn_3$ precipitates. Such nanoscale $Ca_2Mg_6Zn_3$ precipitates not only hinder dislocation movement but also promote dislocation multiplication [32], which facilitates dimple formation and improves the material toughness. Overall, the fracture surface clearly indicates the remarkable ductility of the SR-3ZX alloy.

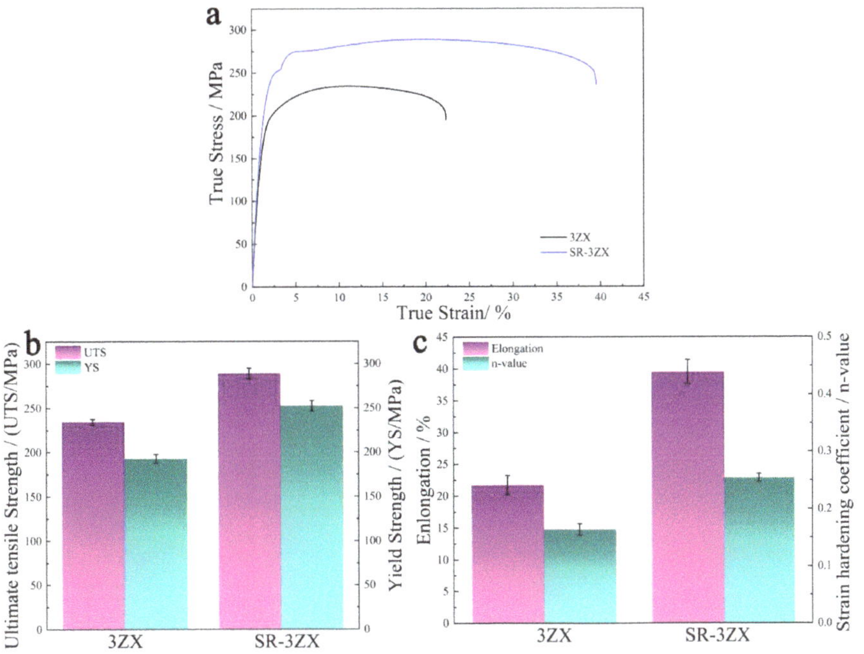

Figure 3. (**a**) Tensile stress–strain curves and the (**b**,**c**) corresponding mechanical properties of 3ZX and SR-3ZX: (**b**) UTS (left axis) and YS (right axis), as well as (**c**) elongation (left axis) and strain hardening coefficient (right axis). The error bars indicate the standard deviation (n = 3).

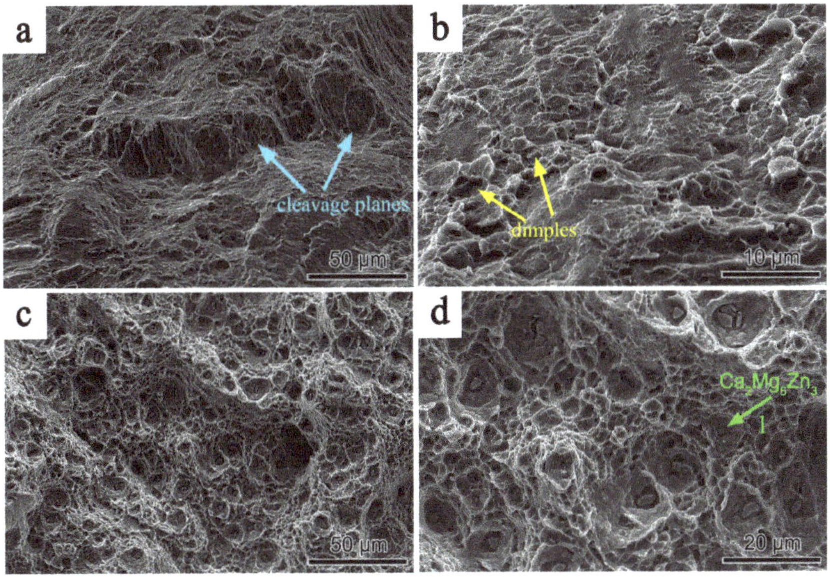

Figure 4. SEM images of fracture surfaces of (**a**,**b**) 3ZX and (**c**,**d**) SR-3ZX.

3.2. Microstructures

Figure 5 shows optical images and transverse cross-sectional SEM images of the microstructures of 3ZX and SR-3ZX. Figure 5a,b reveal that, unlike SR-3ZX, which comprises ultra-fine equiaxed DRX grains, the 3ZX structure contains both the original coarse grains and fine equiaxed grains. This indicates that the shear strain induced by SR provides sufficient energy for DRX. As shown in Figure 5c,d, many precipitates are uniformly distributed in both alloys. By combining the phase diagram with the EDS results from points 1–4 in Figure 5c,d, which are summarised in Table 1, these precipitates are determined to be $Ca_2Mg_6Zn_3$ and Mg_2Zn_{11}.

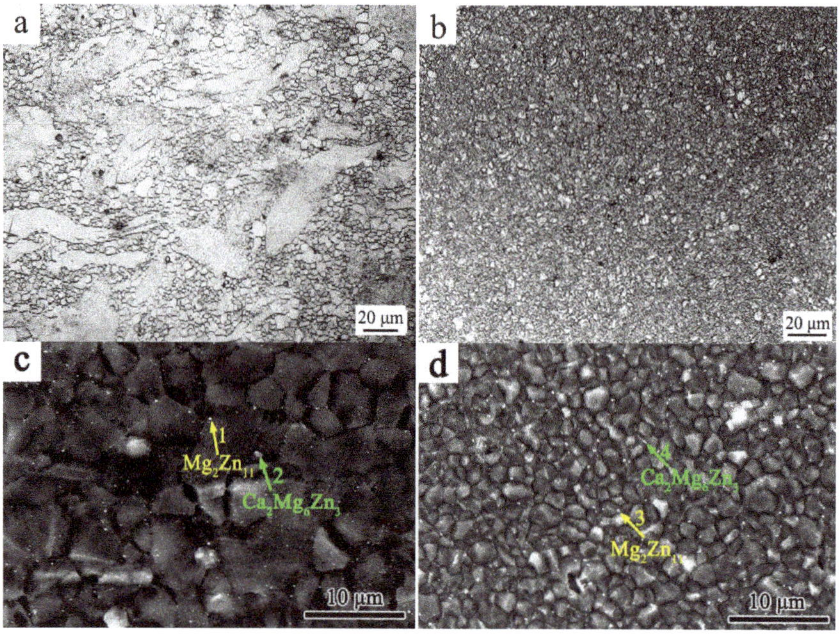

Figure 5. (**a**,**b**) Optical and (**c**,**d**) SEM images of the microstructures of (**a**,**c**) 3ZX and (**b**,**d**) SR-3ZX.

Table 1. Chemical compositions (at.%) of points 1–4 in Figure 5 as determined via EDS.

Location	Mg	Ca	Zn
1	58.66	0.97	40.37
2	64.05	15.03	20.92
3	59.83	1.01	39.16
4	61.82	14.33	23.85

On average, the SR-3ZX grains are significantly smaller and more uniform in size than the 3ZX grains. To further observe the SR-3ZX microstructure, TEM images were obtained (Figure 6). The bright-field (BF) TEM image of SR-3ZX in Figure 6a displays lattice distortion within the grains, which appears black in this image. The high-angle annular dark-field (HAADF) image of SR-3ZX in Figure 6b reveals a large amount of nanoscale precipitates with an average diameter of ~34 nm. The average interparticle spacing was determined to be ~338 nm by the linear intercept method. In SR-3ZX, these precipitates are evenly spread throughout both the grain boundaries and the interiors of the grains. Selected area electron diffraction (SAED) patterns were recorded from points A–C in Figure 6c, which identified the phases as α-Mg, Mg_2Zn_{11}, and $Ca_2Mg_6Zn_3$, respectively (Figure 6d–f). These findings are consistent with the SEM results in Figure 5.

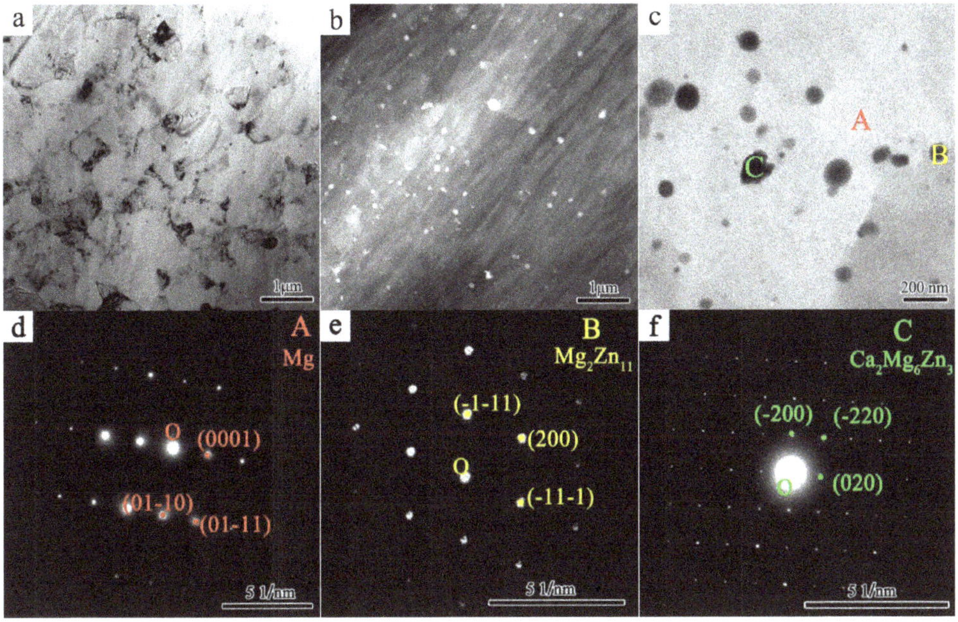

Figure 6. TEM images of SR-3ZX. (**a**) Bright-field (BF), (**b–d**) HAADF, and (**c**) precipitate phase images; (**d–f**) diffraction spots of points A–C in (**c**).

Figure 7 shows the EBSD results for 3ZX and SR-3ZX. The inverse pole figure (IPF) maps of 3ZX and SR-3ZX in Figure 7a,b, respectively, reveal that SR reconfigures the grain boundaries. The 3ZX specimen exhibits a distinct bimodal microstructure comprising equiaxed recrystallised grains with diameters of several micrometres and non-recrystallised regions with coarse elongated grains. By contrast, the SR-3ZX sample is fully recrystallised. In addition, the grains are more randomly orientated in the SR sample, as evidenced by the more-randomly coloured grains in its IPF (Figure 7a). Furthermore, as shown in Figure 7e,f, the average grain sizes of 3ZX and SR-3ZX are 32.5 and 1.6 μm, respectively, corresponding to a reduction of 95.1% after SR. The corresponding kernel average misorientation (KAM) map in Figure 7c indicates that, in the non-recrystallised region of 3ZX, more severe lattice distortion is retained than in the recrystallised region; this contributes to a high geometrically necessary dislocation (GND) density of 1.21×10^{14} m^{-2}, which was calculated using ATEX software based on the EBSD data. For SR-3ZX (Figure 7d), however, the residual stress distribution is more uniform, and the high-stress regions are smaller. Further, the GND density decreases to 5.35×10^{13} m^{-2}. This indicates that the internal stress and high density of dislocations in the material are released via recrystallisation during the SR process, making the internal stress in SR-3ZX lower than that of 3ZX. The IPF for 3ZX (Figure 7g) reveals a bimodal texture (similar to a rare-earth texture), indicating the formation of a non-basal texture during the hot extrusion procedure. Following SR, however, the grain boundary misorientation angle of SR-3ZX (Figure 7h) is deflected, forming a similar <10−12>//RD texture [33,34]. Compared with that of 3ZX, the maximum texture intensity of SR-3ZX decreases from 7.80 to 1.99, implying superior plasticity.

Figure 8 shows the Schmid factors (SFs) of different slip systems in 3ZX and SR-3ZX. In 3ZX, the SF of the basal <*a*> slip is lower than those of the prismatic <*a*>, pyramidal <*a*>, and pyramidal <*a* + *c*> slips, indicating that the activation of the non-basal slip systems is more likely. By contrast, the SFs of the different slip systems in SR-3ZX do not differ significantly, indicating that more slip systems may have been activated. In addition, the average SF of the basal slip increases after SR, which improves the material ductility.

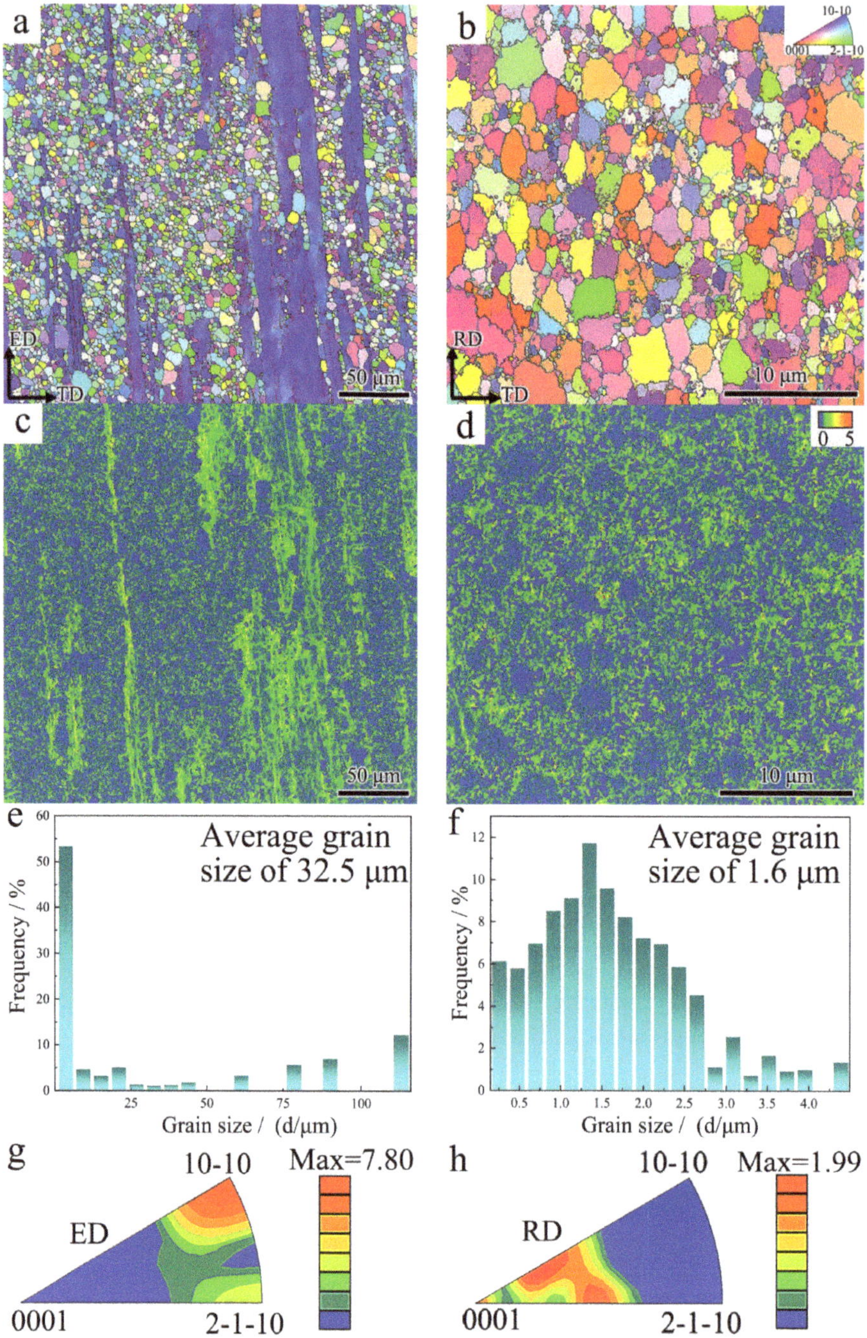

Figure 7. EBSD results for 3ZX (**a**,**c**,**e**,**g**) and SR-3ZX (**b**,**d**,**f**,**h**): (**a**,**b**) IPF maps, (**c**,**d**) KAM maps, (**e**,**f**) grain size distributions, and (**g**,**h**) IPFs.

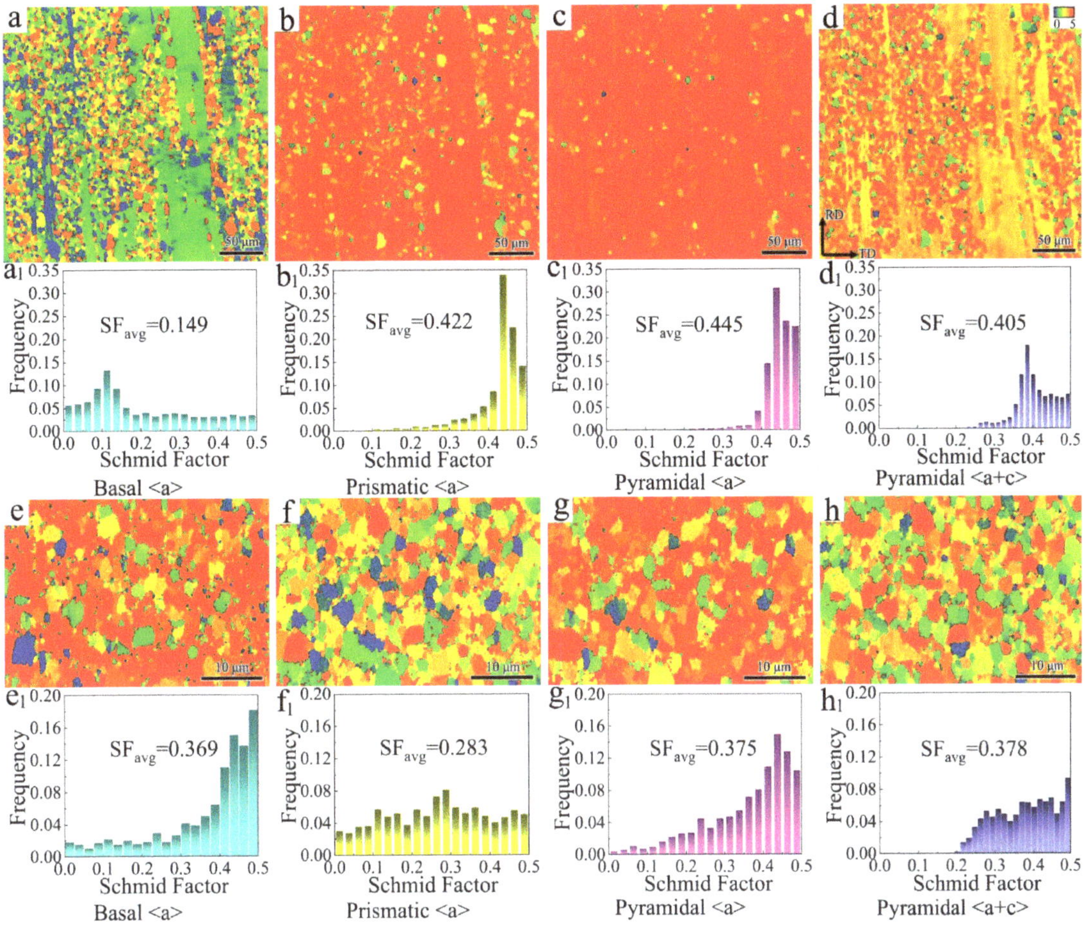

Figure 8. SF maps for various slip systems in 3ZX (**a–d**) and SR-3ZX (**e–h**) (basal: {0 0 0 1}<1 1 −2 0>, prismatic: {1 0 −1 0}<1 1 −2 0>, pyramidal: {1 0 −1 1}<1 1 −2 0>, pyramidal: {1 1 −2 2}<1 1 −2 3>), and corresponding histograms illustrated in (a1–h1), respectively.

4. Discussion

Figure 9 presents a comparison of the mechanical properties of the materials assessed in this study with those of other biomedical Mg–Zn alloys [35–41]. Notably, the SR-3ZX alloy produced in this work is considerably more ductile than all the other Mg–Zn alloys (Figure 9a), but its strength is comparable to many of the high-strength Mg–Zn alloys (Figure 9b). Thus, the SR processing of these alloys overcomes the strength–ductility trade-off found in previously documented Mg–Zn alloys. The elevated ductility indicates the enhanced formability of the Mg alloy at ambient temperature, highlighting the significant potential of SR-3ZX as a Mg-based biodegradable clip.

During SR plastic deformation, the bar experiences uniform stress in three directions and moves forward while spiralling due to the frictional force between the roll and the bar. Throughout several passes, the bar undergoes torsion deformation, leading to a decrease in diameter and an increase in length. As deformation occurs, the external shear and compressive stresses gradually infiltrate the metal bar, causing metal flow and plastic deformation, which subsequently refine the grains within the microstructure. In particular, the strain energy generated by SR encourages grain nucleation for DRX, leading to notable grain refinement of SR-3ZX under the influence of dual-stress conditions.

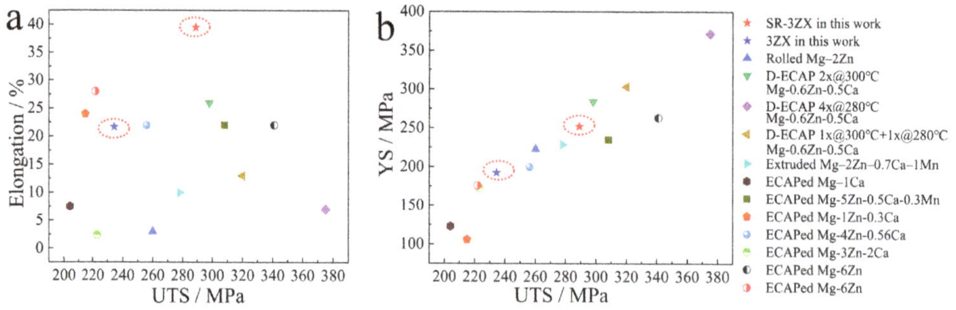

Figure 9. Comparison of the mechanical properties of other biomedical Mg–Zn alloys with those prepared in this study: (**a**) Values of UTS and elongation, and (**b**) values of UTS and YS.

The influence of grain refinement on the YS can be assessed through the Hall–Petch relationship. In addition, residual dislocations can enhance the strength. The presence of GNDs impedes dislocation activation during plastic deformation, thereby enhancing the YS of SR-3ZX. Furthermore, the Orowan mechanism, which refers to dislocations bowing around obstacles such as precipitates, can also strengthen the material. In SR-3ZX, the high number fraction of nanoscale secondary phases and the short distance between these precipitates substantially increased the degree of Orowan strengthening.

To further quantify the contributions of the grain boundary strengthening (σ_{GB}), Orowan strengthening (σ_{Orowan}), and dislocation strengthening ($\sigma_{Dislocation}$) mechanisms to the YS of SR-3ZX, the theoretical YS value of each possible strengthening mechanism was calculated using the following equations [42–48]:

$$\sigma_{GB} = \sigma_0 + kd^{-1/2}, \tag{1}$$

$$\sigma_{Orowan} = \frac{0.4MGb}{\pi\lambda}\ln\frac{d1}{b} / \sqrt{1-\nu_{Mg}}, \tag{2}$$

$$\sigma_{Dislocation} = M\alpha Gb\sqrt{\rho_{Dislocation}}, \tag{3}$$

where σ_0 is 44 MPa for the Mg–Zn–Ca alloy [13]; k represents the slope of the Hall–Petch relation, with a value of 153 MPa $\mu m^{1/2}$; d refers to the average grain size of SR-3ZX; the value of the Taylor factor, symbolized as M, is 3.5; the shear modulus, indicated by G, is 17 GPa for the Mg–Zn–Ca alloy; the Burgers vector, symbolized as b, has a magnitude of 0.32 nm; λ refers to the average interparticle spacing; and the effective diameter of the precipitates is indicated by d_1. For magnesium, Poisson's ratio, denoted as ν_{Mg}, is 0.35. The numerical constant α, which is 0.2, assesses the efficiency of dislocation strengthening. The total dislocation density, represented by $\rho_{Dislocation}$, is approximately 1.5 times greater than the GND density.

The contributions of σ_{GB}, σ_{Orowan}, and $\sigma_{Dislocation}$ to the YS of SR-3ZX were calculated to be 165.0, 41.5, and 34.1 MPa, respectively. The contributions of the various strengthening mechanisms and a comparison with the experimental results are shown in Figure 10. The predicted YS was 240.6 MPa, which is slightly lower than the experimental YS value of ~252.3 MPa. This discrepancy can possibly be attributed to texture strengthening, as the texture was not high and therefore was not considered in the analysis. Nonetheless, the three aforementioned strengthening mechanisms could be concluded to synergistically strengthen SR-3ZX [49–51]. Grain refinement was the main mechanism for strengthening in SR-3ZX; however, the contributions of the Orowan and dislocation strengthening mechanisms were also not trivial.

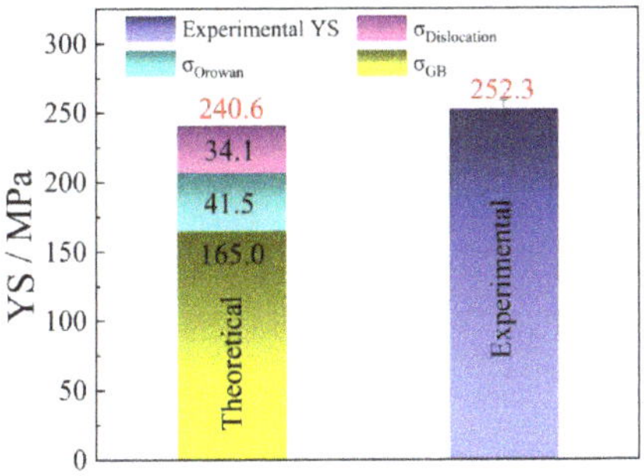

Figure 10. Quantified contributions of different strengthening mechanisms and a comparison of the theoretical YS with the experimentally determined YS of SR-3ZX, where the theoretical YS (σ_{Theory}) = $\sigma_{GB} + \sigma_{Orowan} + \sigma_{Dislocation}$.

Additionally, SR-3ZX showed exceptional ductility, achieving a fracture elongation reaching 39.5%. To explore the properties underlying this exceptional ductility, the TEM results used to identify the activated dislocations of SR-3ZX under the double-beam condition are shown in Figure 11. Figure 11a,b present BF and dark-field (DF) images under the two-beam condition of *g* = [−1011], respectively. Numerous <*a*> dislocations (marked in red) and a small number of <*c*> dislocations (yellow) are clearly identified. The <*c*> and <*c* + *a*> dislocations (green) are observed under *g* = [−110−1] (Figure 11b). As shown in Figure 11c,f, under the *g* = [−2−110] vector, <*a*> dislocations are mainly observed, accompanied by a small number of <*c* + *a*> dislocations. These results indicate that <*a*>, <*c*>, and <*c* + *a*> dislocations can all be activated in SR-3ZX. For HCP metals, the scarcity of slip systems often results in insufficient ductility. Therefore, the high plasticity of SR-3ZX is due to the widespread activation of slip systems. The BF TEM images (Figure 11d–f) reveal that a large number of ultra-fine nanoprecipitates pin the dislocations, and the activation of various slip systems increases the number of dislocation intersections caused by different dislocation movements. These enhance the resistance to dislocation movement through SR-3ZX, thereby improving its strength.

Under SR, the bimodal texture of 3ZX changed to a more uniform <10−12>//RD texture in SR-3ZX because of the large plastic deformation and DRX induced by the SR process. Additionally, the texture weakened. Therefore, the stress concentration could be more easily released, as evidenced by the higher fracture elongation of SR-3ZX. Texture weakening, therefore, contributed to improving the plasticity of SR-3ZX. Moreover, uniformly distributed nanoprecipitates in SR-3ZX (Figure 6) promoted dislocation accumulation, yielding a high ductility [52–54]. In addition, the lower dislocation density in SR-3ZX (Figure 7d) provided additional space for dislocation sliding and accumulation during deformation, which may have reduced the work-hardening capability and improved ductility [55,56]. Finally, the initiation of basal and non-basal slip systems provided more dislocations (Figure 11) to promote plastic deformation; thus, the SR-3ZX strain hardening (Figure 3) and ductility were improved by the SR-induced deformation.

This study demonstrated that SR-3ZX has excellent ductility, showing great promise for fabricating biodegradable surgical clips. However, we only studied its mechanical properties; hence, future studies will focus on carefully investigating its corrosion resistance and biocompatibility as well as optimising its structure.

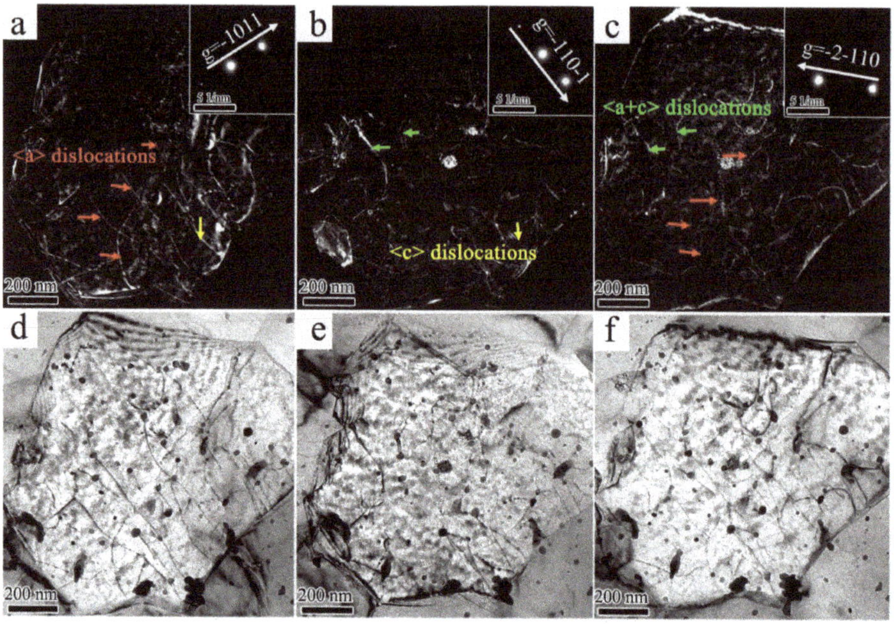

Figure 11. DF (**a**–**c**) and corresponding BF TEM images (**d**–**f**) of SR-3ZX under g = [−1 0 1 1] (**a**,**d**), g = [−1 1 0 −1] (**b**,**e**), and g = [−2 −1 1 0] (**c**,**f**). The red, orange, and green arrows indicate <*a*>, <*c*>, and <*a* + *c*> dislocations.

5. Conclusions

The microstructure, grain refinement, precipitates, texture, and tensile properties of a biomedical Mg–Zn–Ca alloy processed by SR were examined in this study. Under SR, the average grain size decreased significantly from 32.5 μm for 3ZX to 1.6 μm for SR-3ZX. Moreover, SR-3ZX exhibited an ultra-high elongation of ~39.5%, an excellent YS of ~252.3 MPa, and a UTS of ~289 MPa, which were 82.0%, 31%, and 23.3% higher, respectively, than those of 3ZX prior to SR. The high YS of SR-3ZX primarily stemmed from grain boundary strengthening (~165 MPa), with the considerable involvement of Orowan strengthening (~41.5 MPa) induced by precipitates, as well as dislocation strengthening (~34.1 MPa). The weak internal stress, nanoscale precipitates, texture weakening, and activation of multiple slip systems were the main factors generating the ultra-high tensile ductility of SR-3ZX up to ~39.5%. These metrics make SR-3ZX a promising Mg alloy for application in biodegradable clips.

Author Contributions: H.Z.: writing—original draft, data curation, methodology. W.S.: writing—review and editing, investigation. L.D.: formal analysis, investigation. L.Z.: methodology, investigation. K.S.S.: conceptualization, resources. J.Z.: writing—review and editing, supervision. All authors have read and agreed to the published version of the manuscript.

Funding: The authors acknowledge the financial support for this work from the Natural Science Foundation of Shandong Province (Grants No. ZR2023QE119 and ZR2023QE222) and Shandong University of Aeronautics Doctoral Research Initiation Fund Project (No. 2021Y38).

Institutional Review Board Statement: Not Applicable.

Informed Consent Statement: Not Applicable.

Data Availability Statement: The raw/processed data required to reproduce these findings cannot be shared.

Conflicts of Interest: The authors declare no conflict of interest.

References

1. Yu, X.; Li, D.; Liu, Y.; Ding, P.; He, X.; Zhao, Y.; Chen, M.; Liu, D. In vitro and in vivo studies on the degradation and biosafety of Mg-Zn-Ca-Y alloy hemostatic clip with the carotid artery of SD rat model. *Mater. Sci. Eng. C* **2020**, *115*, 111093. [CrossRef] [PubMed]
2. Yoshida, T.; Fukumoto, T.; Urade, T.; Kido, M.; Toyama, H.; Asari, S.; Ajiki, T.; Ikeo, N.; Mukai, T.; Ku, Y. Development of a new biodegradable operative clip made of a magnesium alloy: Evaluation of its safety and tolerability for canine cholecystectomy. *Surgery* **2017**, *161*, 1553–1560. [CrossRef] [PubMed]
3. Li, D.; Zhang, D.; Yuan, Q.; Liu, L.; Li, H.; Xiong, L.; Guo, X.; Yan, Y.; Yu, K.; Dai, Y.; et al. In vitro and in vivo assessment of the effect of biodegradable magnesium alloys on osteogenesis. *Acta Biomater.* **2022**, *141*, 454–465. [CrossRef]
4. Liu, Y.; Liu, L.; Chu, X.; Zhang, Z.; Fu, Z.; Pan, B.; Wang, J.; Tian, H.; Liu, Y.; Yan, Y.; et al. Mechanical properties and bio-corrosion behavior of Mg-Zn-Zr alloy influenced by rotary swaging. *Mater. Today Commun.* **2024**, *39*, 108808. [CrossRef]
5. Peng, B.; Xu, H.; Song, F.; Wen, P.; Tian, Y.; Zheng, Y. Additive manufacturing of porous magnesium alloys for biodegradable orthopedic implants: Process, design, and modification. *J. Mater. Sci. Technol.* **2024**, *182*, 79–110. [CrossRef]
6. Tsakiris, V.; Tardei, C.; Clicinschi, F.M. Biodegradable Mg alloys for orthopedic implants—A review. *J. Magnes. Alloys* **2021**, *9*, 1884–1905. [CrossRef]
7. Zhao, T.; Zhang, Q.; Wang, R.; Wu, J.; Hu, Y.; Jiang, B.; Pan, F. Achieving high strength-ductility synergy in Mg-Gd-Zn-Zr alloy by controlling extrusion processes parameters. *Mater. Sci. Eng. A* **2024**, *916*, 147342. [CrossRef]
8. Zhang, X.; He, Y.; Chen, T.; Bi, G.; Li, Y.; Tang, D.; Wang, X. Coordinating the deformation of a low-alloyed magnesium alloy for a superior combination of strength and ductility through core-shell structured reinforcements. *Acta Mater.* **2024**, *281*, 120365. [CrossRef]
9. Ikeo, N.; Nakamura, R.; Naka, K.; Hashimoto, T.; Yoshida, T.; Urade, T.; Fukushima, K.; Yabuuchi, H.; Fukumoto, T.; Ku, Y.; et al. Fabrication of a magnesium alloy with excellent ductility for biodegradable clips. *Acta Biomater.* **2016**, *29*, 468–476. [CrossRef]
10. Holweg, P.; Berger, L.; Cihova, M.; Donohue, N.; Clement, B.; Schwarze, U.; Sommer, N.G.; Hohenberger, G.; van den Beucken, J.J.J.P.; Seibert, F.; et al. A lean magnesium–zinc–calcium alloy ZX00 used for bone fracture stabilization in a large growing-animal model. *Acta Biomater.* **2020**, *113*, 646–659. [CrossRef]
11. Cho, D.H.; Avey, T.; Nam, K.H.; Dean, D.; Luo, A.A. In vitro and in vivo assessment of squeeze-cast Mg-Zn-Ca-Mn alloys for biomedical applications. *Acta Biomater.* **2022**, *150*, 442–455. [CrossRef] [PubMed]
12. Chen, Y.; Xu, Z.; Smith, C.; Sankar, J. Recent advances on the development of magnesium alloys for biodegradable implants. *Acta Biomater.* **2014**, *10*, 4561–4573. [CrossRef]
13. Zhang, H.; Ding, Y.; Li, R.; Gao, Y. Enhanced strength-ductility synergy and activation of non-basal slip in as-extruded Mg–Zn–Ca alloy via heterostructure. *J. Mater. Res. Technol.* **2024**, *28*, 1841–1851. [CrossRef]
14. Zhang, B.P.; Geng, L.; Huang, L.J.; Zhang, X.X.; Dong, C.C. Enhanced mechanical properties in fine-grained Mg-1.0Zn-0.5Ca alloys prepared by extrusion at different temperatures. *Scr. Mater.* **2010**, *63*, 1024–1027. [CrossRef]
15. Horky, J.; Bryta, K.; Krystian, M.; Mozdzen, G.; Mingler, B.; Sajti, L. Improving mechanical properties of lean Mg–Zn–Ca alloy for absorbable implants via Double Equal Channel Angular Pressing (D-ECAP). *Mater. Sci. Eng. A* **2021**, *826*, 142002. [CrossRef]
16. Mao, L.; Zhou, Y.; Zheng, X.; Cai, X.; Chen, Y.; Yang, W.; Wang, J.; Zhang, J.; Song, C. Structural optimization and in vitro corrosion analysis of biodegradable Mg-Nd-Zn-Zr alloy clip. *J. Mech. Behav. Biomed. Mater.* **2025**, *161*, 106790. [CrossRef] [PubMed]
17. Han, Y.; Zhang, X.B.; Wang, Z.X.; Xiao, Y.; Gao, X.L.; Zhao, Y.Q. Study on cooling process of copper tube after three-roll planetary rolling. *Int. Commun. Heat Mass* **2020**, *110*, 104393. [CrossRef]
18. Shih, C.K.; Hung, C.; Hsu, R.Q. The fnite element analysis on planetary rolling process. *J. Mater. Process. Technol.* **2001**, *113*, 115–123. [CrossRef]
19. Shih, C.K.; Hung, C. Experimental and numerical analyses on three-roll planetary rolling process. *J. Mater. Process. Technol.* **2003**, *142*, 702–709. [CrossRef]
20. Komori, K. Simulation of deformation and temperaturein multi-pass three-roll rolling. *J. Mater. Process. Technol.* **1999**, *92–93*, 450–457. [CrossRef]
21. Ji, C.; Niu, H.; Li, Z.; Wang, T.; Huang, Q. Deformation law and bonding mechanism of 45 carbon steel/316L stainless steel cladding tubes fabricated by three-roll skew rolling bonding process. *J. Mater. Process. Technol.* **2024**, *325*, 118277. [CrossRef]
22. Hwang, Y.M.; Tsai, W.M.; Tsai, F.H.; Her, I. Analytical and experimental study on the spiral marks of the rolled product during three-roll planetary rolling processes. *Int. J. Mach. Tools Manuf.* **2006**, *46*, 1555–1562. [CrossRef]
23. Wang, Y.L.; Molotnikov, A.; Diez, M.; Lapovok, R.; Kim, H.-E.; Wang, J.T.; Estrin, Y. Gradient structure produced by three roll planetary milling: Numerical simulation and microstructural observations. *Mater. Sci. Eng. A* **2015**, *639*, 165–172. [CrossRef]
24. Diez, M.; Kim, H.E.; Serebryany, V.; Dobatkin, S.; Estrin, Y. Improving the mechanical properties of pure magnesium by three-roll planetary milling. *Mater. Sci. Eng. A* **2014**, *612*, 287–292. [CrossRef]

25. Bahmani, A.; Arthanari, S.; Shin, K.S. Improvement of corrosion resistance and mechanical properties of a magnesium alloy using screw rolling. *J. Alloys Compd.* **2020**, *813*, 152155. [CrossRef]
26. Wang, L.; Zhang, Z.; Zhang, H.; Wang, H.; Shin, K.S. The dynamic recrystallization and mechanical property responses during hot screw rolling on pre-aged ZM61 magnesium alloys. *Mater. Sci. Eng. A* **2020**, *798*, 140126. [CrossRef]
27. Ding, P.; Liu, Y.; He, Y.; Liu, D.; Chen, M. In vitro and in vivo biocompatibility of Mg-Zn-Ca alloy operative clip. *Bioact. Mater.* **2019**, *4*, 236–244. [CrossRef]
28. Yang, M.; Liu, D.; Zhang, R.; Chen, M. Microstructure and Properties of Mg-3Zn-0.2Ca Alloy for Biomedical Application. *Rare Met. Mater. Eng.* **2018**, *47*, 0093–0098. [CrossRef]
29. Zheng, H.; Chen, M.; Li, Z.; Yu, L. Mechanism study on improving the corrosion resistance of screw rolled Mg–Zn–Ca alloy by nano-MgO addition. *J. Mater. Res. Technol.* **2024**, *31*, 3104–3116. [CrossRef]
30. Wu, L.; Jain, A.; Brown, D.; Stoica, G.; Agnew, S.; Clausen, B.; Fielden, D.; Liaw, P. Twinning-detwinning behavior during the strain-controlled low-cycle fatigue testing of a wrought magnesium alloy, ZK60A. *Acta Mater.* **2008**, *56*, 688–695. [CrossRef]
31. Song, B.; Xin, R.; Chen, G.; Zhang, X.; Liu, Q. Improving tensile and compressive properties of magnesium alloy plates by pre-cold rolling. *Scr. Mater.* **2012**, *66*, 1061–1064. [CrossRef]
32. Yang, H.; Li, K.; Bu, Y.; Wu, J.; Fang, Y.; Meng, L.; Liu, J.; Wang, H. Nanoprecipitates induced dislocation pinning and multiplication strategy for designing high strength, plasticity and conductivity Cu alloys. *Scr. Mater.* **2021**, *195*, 113741. [CrossRef]
33. Pu, Z.; Song, G.-L.; Yang, S.; Outeiro, J.; Dillon, O.; Puleo, D.; Jawahir, I. Grain refined and basal textured surface produced by burnishing for improved corrosion performance of AZ31B Mg alloy. *Corros. Sci.* **2012**, *57*, 192–201. [CrossRef]
34. Jiang, M.G.; Xu, C.; Yan, H.; Lu, S.H.; Nakata, T.; Lao, C.S.; Chen, R.S.; Kamado, S.; Han, E.H. Correlation between dynamic recrystallization and formation of rare earth texture in a Mg-Zn-Gd magnesium alloy during extrusion. *Sci. Rep.* **2018**, *8*, 16800. [CrossRef]
35. Bazhenov, V.; Li, A.; Komissarov, A.; Koltygin, A.; Tavolzhanskii, S.; Bautin, V.; Voropaeva, O.; Mukhametshina, A.; Tokar, A. Microstructure and mechanical and corrosion properties of hot-extruded Mg–Zn–Ca–(Mn) biodegradable alloys. *J. Magnes. Alloys* **2021**, *9*, 1428–1442. [CrossRef]
36. Doležal, P.; Zapletal, J.; Fintová, S.; Trojanová, Z.; Greger, M.; Roupcová, P.; Podrábský, T. Influence of processing techniques on microstructure and mechanical properties of a biodegradable Mg-3Zn-2Ca alloy. *Materials* **2016**, *9*, 880. [CrossRef]
37. Li, W.; Shen, Y.; Shen, J.; Shen, D.; Liu, X.; Zheng, Y.; Yeung, K.W.; Guan, S.; Kulyasova, O.B.; Valiev, R. In vitro and in vivo studies on pure Mg, Mg–1Ca and Mg–2Sr alloys processed by equal channel angular pressing. *Nano Mater. Sci.* **2020**, *2*, 96–108. [CrossRef]
38. Vinogradov, A.; Vasilev, E.; Kopylov, V.I.; Linderov, M.; Brilevesky, A.; Merson, D. High performance fine-grained biodegradable Mg-Zn-Ca alloys processed by severe plastic deformation. *Metals* **2019**, *9*, 186. [CrossRef]
39. Tong, L.B.; Chu, J.H.; Jiang, Z.H.; Kamado, S.; Zheng, M.Y. Ultra-fne grained Mg-Zn-Ca-Mn alloy with simultaneously improved strength and ductility processed by equal channel angular pressing. *J. Alloys Compd.* **2019**, *785*, 410–421. [CrossRef]
40. Wang, W.; Blawert, C.; Zan, R.; Sun, Y.; Peng, H.; Ni, J.; Han, P.; Suo, T.; Song, Y.; Zhang, S.; et al. A novel lean alloy of biodegradable Mg–2Zn with nanograins. *Bioact. Mater.* **2021**, *6*, 4333–4341. [CrossRef]
41. Yan, J.; Qin, Z.; Yan, K. Mechanical properties and microstructure evolution of Mg-6 wt % Zn alloy during equal-channel angular pressing. *Metals* **2018**, *8*, 841. [CrossRef]
42. Bian, M.; Huang, X.; Chino, Y. A room temperature formable magnesium–silver–calcium sheet alloy with high ductility. *Mater. Sci. Eng. A* **2020**, *774*, 138923. [CrossRef]
43. Gao, W.; Wang, X.; Lin, Y.; Wang, X.; Liu, D.; Sun, X. Achieving ultra-high strength and ductility in a rare-earth-free magnesium alloy via precisely controlled secondary hot extrusion process with an extremely low extrusion speed. *J. Magnes. Alloys* **2024**, *12*, 5216–5230. [CrossRef]
44. Wong, W.L.E.; Gupta, M. Development of Mg/Cu nanocomposites using microwave assisted rapid sintering. *Compos. Sci. Technol.* **2007**, *67*, 1541–1552. [CrossRef]
45. Habibi, M.K.; Joshi, S.P.; Gupta, M. Hierarchical magnesium nano-composites for enhanced mechanical response. *Acta Mater.* **2010**, *58*, 6104–6114. [CrossRef]
46. Han, B.Q.; Dunand, D.C. Microstructure and mechanical properties of magnesium containing high volume fractions of yttria dispersoids. *Mater. Sci. Eng. A* **2000**, *277*, 297–304. [CrossRef]
47. Kang, Y.C.; Chan, S.L.I. Tensile properties of nanometric Al_2O_3 particulate-reinforced aluminum matrix composites. *Mater. Chem. Phys.* **2004**, *85*, 438–443. [CrossRef]
48. Cáceres, C.H.; Lukác, P. Strain hardening behaviour and the Taylor factor of pure magnesium. *Philos. Mag.* **2008**, *88*, 977–989. [CrossRef]
49. Shahin, M.; Munir, K.; Wen, C.; Li, Y. Magnesium matrix nanocomposites for orthopedic applications: A review from mechanical, corrosion, and biological perspectives. *Acta Biomater.* **2019**, *96*, 1–19. [CrossRef]

50. Naghdi, F.; Mahmudi, R.; Kang, J.Y.; Kim, H.S. Contributions of different strengthening mechanisms to the shear strength of an extruded Mg-4Zn-0.5Ca alloy. *Philos. Mag.* **2015**, *95*, 3452–3466. [CrossRef]
51. Habibnejad-Korayem, M.; Mahmudi, R.; Poole, W.J. Enhanced properties of Mg-based nano-composites reinforced with Al_2O_3 nano-particles. *Mater. Sci. Eng. A* **2009**, *519*, 198–203. [CrossRef]
52. Soleimani, M.; Kalhor, A.; Mirzadeh, H. Transformation-induced plasticity (TRIP) in advanced steels: A review. *Mater. Sci. Eng. A* **2020**, *795*, 140023. [CrossRef]
53. Gao, Q.; Wei, R.; Feng, S.; Chen, C.; Han, Z.; Chen, L.; Wang, T.; Wu, S.; Li, F. Ultrahigh cryogenic strength and ductility in a duplex metastable ferrous medium-entropy alloy. *Scr. Mater.* **2023**, *228*, 115334. [CrossRef]
54. Zhang, Y.; Ding, H. Ultrafine also can be ductile: On the essence of Lüders band elongation in ultrafine-grained medium manganese steel. *Mater. Sci. Eng. A* **2018**, *733*, 220–223. [CrossRef]
55. Wang, N.; Chen, Y.; Wu, G.; Zhao, Q.; Zhang, Z.; Zhu, L.; Luo, J. Non-equivalence contribution of geometrically necessary dislocation and statistically stored dislocation in work-hardened metals. *Mater. Sci. Eng. A* **2022**, *836*, 142728. [CrossRef]
56. Zhang, H.; Wang, H.; Wang, J.; Rong, J.; Zha, M.; Wang, C.; Ma, P.; Jiang, Q. The synergy effect of fine and coarse grains on enhanced ductility of bimodal-structured Mg alloys. *J. Alloys Compd.* **2019**, *780*, 312–317. [CrossRef]

Disclaimer/Publisher's Note: The statements, opinions and data contained in all publications are solely those of the individual author(s) and contributor(s) and not of MDPI and/or the editor(s). MDPI and/or the editor(s) disclaim responsibility for any injury to people or property resulting from any ideas, methods, instructions or products referred to in the content.

materials

MDPI

Article

Influence of Laser Power on CoCrFeNiMo High-Entropy Alloy Coating Microstructure and Properties

Shuai Li [1,2,*], Fuheng Nie [1,3,*], Jiyuan Ding [2], Guijun Mao [1,3], Yang Guo [1,3], Tianlan Cao [1,3], Chong Xiang [1,3] and Honggang Dong [4]

1 State Key Laboratory of Clean and Efficient Turbomachinery Power Equipment, Deyang 618000, China; maoguijun0211@163.com (G.M.); guoyang0192@163.com (Y.G.); caotianlan333@163.com (T.C.); xiangchong0311@163.com (C.X.)
2 School of Mechanical Engineering, North China University of Water Resources and Electric Power, Zhengzhou 450045, China; dingjiyuan65959610@163.com
3 Dongfang Electric Corporation Dongfang Turbine Co., Ltd., Deyang 618000, China
4 School of Materials Science and Engineering, Dalian University of Technology, Dalian 116024, China; donghg@dlut.edu.cn
* Correspondence: lyctlishuai@163.com (S.L.); niefuheng0508@163.com (F.N.); Tel.: +86-0371-69127295 (S.L. & F.N.)

Abstract: This work studies the fabrication of CoCrFeNiMo high-entropy alloy (HEA) coatings via coaxial powder-fed laser cladding, addressing porosity and impurity issues in conventional methods. The HEA coatings exhibited eutectic/hypereutectic microstructures under all laser power conditions. A systematic investigation of laser power effects (1750–2500 W) reveals that 2250 W optimizes microstructure and performance, yielding a dual-phase structure with FCC matrix and dispersed σ phases (Fe-Cr/Mo-rich). The coating achieves exceptional hardness (738.3 $HV_{0.2}$, 3.8× substrate), ultralow wear rate (4.55 × 10^{-5} $mm^3/N{\cdot}m$), and minimized corrosion current (2.31 × 10^{-4} A/cm^2) in 3.5 wt.% NaCl. The friction mechanism of the CoCrFeNiMo HEA coating is that in high-speed friction and wear, the oxide film is formed on the surface of the coating, and then the rupture of the oxide film leads to adhesive wear and abrasive wear. The corrosion mechanism is the galvanic corrosion caused by the potential difference between the FCC phase and the σ phase.

Keywords: high-entropy alloy coatings; coaxial powder-feeding laser cladding; CoCrFeNiMo; laser power

1. Introduction

High-entropy alloys (HEAs), an emerging category of metallic materials, are characterized by their unique composition; they contain five or more principal elements blended in approximately equal molar proportions. The highly disordered atomic arrangement induces a transition from ordered to random atomic distribution, resulting in four characteristic effects: thermodynamic high-entropy effect, kinetic sluggish diffusion effect, structural lattice distortion effect, and performance-related "cocktail effect" [1]. As an emerging alloy system, HEAs have demonstrated exceptional properties and promising potential in coating applications [2], positioning them at the forefront of global materials research.

HEA coating technologies have emerged as a research focus in materials science due to their exceptional properties, with multiple fabrication processes being developed to date [3]. Based on distinct processing principles, these techniques can be primarily categorized into three groups: 1. deposition methods (e.g., electrochemical deposition [4] and physical vapor deposition (PVD [5])); 2. spraying techniques (including HVOF spraying [6], cold spraying [7], and plasma spraying processes [8,9]); and 3. cladding techniques (encompassing plasma cladding [10,11], gas tungsten arc cladding [12], electron beam cladding [13], and laser cladding [14–16]). As shown in Figure 1, among these, laser cladding has demonstrated unique advantages through its high energy density (typically 10^3–10^4 W/mm^2) and rapid cooling rates (10^3–10^6 K/s), enabling the fabrication of coatings with superior characteristics: low dilution rates, narrow heat-affected zones (HAZ < 200 μm), and excellent metallurgical bonding (interface strength > 400 MPa) [17]. These attributes have rendered it particularly advantageous for critical applications in marine engineering and precision machinery manufacturing.

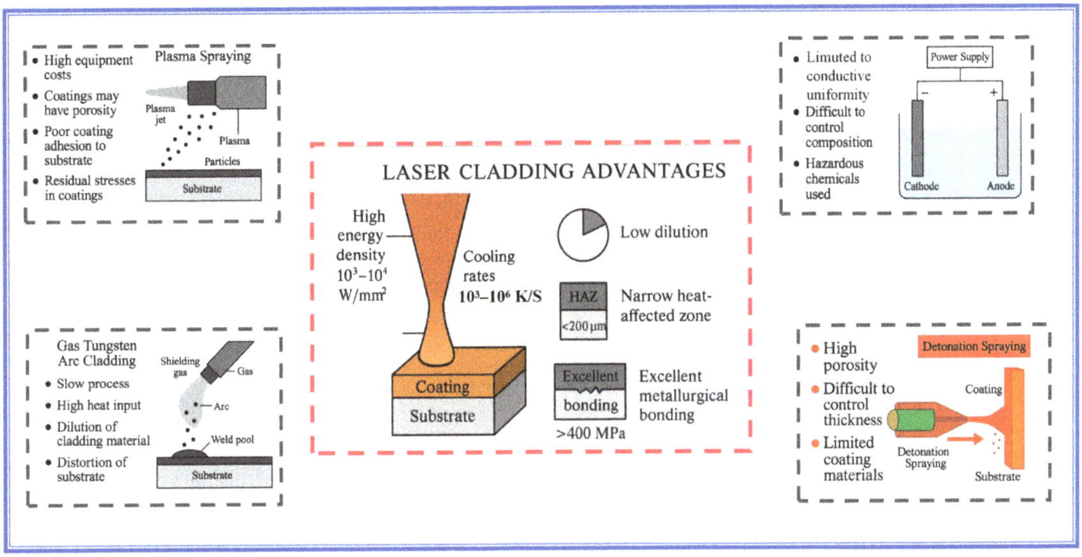

Figure 1. The advantages of laser cladding.

Current research in laser cladding coating fabrication predominantly employs a two-stage methodology: initial powder preplacement onto the substrate surface, followed by laser irradiation to achieve simultaneous fusion of the deposited material with the base substrate, thereby forming high-quality coatings. Zhang et al. [18] synthesized CoCrFeNiTi HEA coatings on 304SS via laser cladding using ball-milled pre-alloyed powders. The coating microstructure comprises an FCC solid-solution matrix and Fe-Cr-enriched Laves phase precipitates, accompanied by directionally solidified columnar grains and eutectic morphology. Dual strengthening mechanisms (solid-solution and second-phase reinforcement) achieved 2.4 times the substrate hardness (568 $HV_{0.2}$) and a lower erosion wear rate, demonstrating exceptional wear resistance. Zeng et al. [19] prepared a laser-clad coating on cleaned Ti-6Al-4V substrates by pre-depositing a 2 mm layer of ball-milled blended Al/Co/Cr/Fe/Ni/Ti powders. The coating exhibited a dual-phase microstructure: BCC phases predominantly located in dendritic cores and FCC phases concentrated in interdendritic regions. Oxidation tests revealed exceptional performance with a parabolic rate constant tenfold lower than that of the substrate. High-temperature tribological evaluation at 600 °C demonstrated a wear volume loss of 4.97×10^{-4} mm^3/N·m, representing a 13-fold reduction compared to the substrate. Zhang et al. [20] fabricated laser-clad coatings on Q235 steel using Fe/Co/Ni/Cr/Cu pre-mixed powders with individual additions of trace Si, Mn, and Mo elements. The coatings maintained a single FCC structure, while elemental doping significantly enhanced microhardness, elevated softening onset temperature, and reduced corrosion current density through synergistic solid-solution strengthening and passive film optimization. Currently, most research on laser cladding HEA coatings focuses on pre-coating the substrate with powder, followed by scanning with a high-energy laser beam to form the coating, but this method relies on the use of organic solvents. This can lead to the formation of pores and the introduction of impurities during the cooling process of the laser melt pool. In contrast, coaxial powder-fed laser cladding technology is suitable for surfaces with complex geometries and allows for precise control of the cladding area without the need for binders. This eliminates issues related to porosity and impurities while offering broader application potential.

The transition metal elements Co, Cr, Fe, and Ni are conventionally selected as primary constituents of HEAs due to their comparable atomic radii. However, FeCoCrNi HEA coatings predominantly form single-phase FCC structures, whose low hardness limits their applicability in high-stress environments [21]. To address this, researchers have opted to introduce additional principal elements to engineer phase-structured HEAs with distinct physicochemical properties. For instance, Al additions induce dual-phase solid solution microstructures via FCC to BCC phase transformation mechanisms [22]. Nevertheless, single solid-solution HEAs still demonstrate suboptimal performance under complex tribological and corrosive service conditions.

Mo has gained prominence for its capacity to form intermetallic compounds with exceptional physical properties, mechanical robustness, and thermal stability. Incorporating Mo into quaternary FeCoCrNi HEAs promotes the precipitation of topologically close-packed (TCP) phases (e.g., σ phase), which generate residual compressive stresses to inhibit crack propagation and enhance fatigue resistance [23]. The synergistic coexistence of FCC matrices and hard intermetallic compounds endows these coatings with superior comprehensive mechanical properties, enabling exceptional performance in multifactorial service environments involving wear–corrosion interactions [24]. Chang et al. [25] mechanically mixed different proportions of Co, Cr, Fe, Ni, and Mo powders to obtain $CoCrFeNiMo_x$ (x = 0, 0.5, 1.0, 1.5) mixed powders. After blending the powders with an organic solvent, they were pre-coated on the surface of AISI1045 steel to prepare laser cladding HEA coatings. The $CoCrFeNiMo_x$ coatings are primarily composed of an FCC phase. The addition of Mo increases the lattice distortion of the FCC matrix. When the Mo content is high ($x \geq 1.0$), a small amount of BCC phase or σ phase appears, and the microstructure transitions from columnar grains to equiaxed grains. At higher Mo contents, grain refinement occurs. $CoCrFeNiMo_{1.5}$ demonstrates superior hardness and exceptional wear resistance, primarily attributed to three synergistic mechanisms: solid-solution strengthening, refined grain structures, and precipitation hardening induced by the σ phase enriched with Fe, Cr, and Mo. Zeng et al. [26] used ball-milled Co, Cr, Fe, Ni, Mo, Ti, and Mn powders to prepare coatings on the surface of low-carbon steel through laser cladding. The CoCrFeNi and CoCrFeNiMn HEA coatings exhibit single-phase FCC structures, while the CoCrFeNiTi coating comprises FCC, Laves, and η phases, and the CoCrFeNiMo coating consists of FCC and σ phases. Among these, the CoCrFeNiMo coating demonstrates the highest hardness (~700 HV) and optimal wear resistance due to σ-phase strengthening, with oxidative wear identified as the dominant wear mechanism. However, the multiphase structures induce galvanic corrosion effects, resulting in comparatively inferior corrosion resistance for both CoCrFeNiMo and CoCrFeNiTi coatings.

Among various HEA systems, the CoCrFeNiMo series has been particularly noteworthy for its dual-phase structure and has proven to be a high-performance material in diverse fields, including marine engineering, aerospace, and high-speed rail applications [27]. Currently, there has been no research on the preparation of CoCrFeNiMo HEA coatings using coaxial powder-fed laser cladding, nor on the effect of laser cladding parameters on the performance of the coatings. In the process of fabricating CoCrFeNiMo HEA coatings via coaxial powder-feeding laser cladding, laser power plays a critical regulatory role in the microstructural evolution and macroscopic performance of the coatings. Notably, while existing studies predominantly focus on multi-parameter coupling effects, there is still a gap in the systematic study of the coating structure evolution, tribological properties, and electrochemical corrosion mechanisms under univariate regulation strategies (such as laser power gradient changes). This study employed coaxial powder-feeding laser cladding to deposit CoCrFeNiMo coatings on Q355 steel substrates. By systematically varying four laser power parameters, this study elucidated the influence of laser energy density on microstructure and properties.

2. Materials and Methods

2.1. Coating Deposition

The experimental platform employed Q355 steel substrates (100 mm × 100 mm × 20 mm). The Q355 steel used complies with ATSM A992 [28], and the quantitative characterization of its elemental composition is shown in Table 1. As the cladding material, CoCrFeNiMo HEA powder was procured from Avimetal Powder Metallurgy Technology Co., Ltd. (Beijing, China). The physical appearance of the HEA powder is depicted in Figure 2b.

Table 1. Chemical composition of Q355 steel (wt.%).

C	Si	Mn	P	S	Cr	Ni	Cu	Fe
≤0.24	≤0.55	≤1.60	≤0.035	≤0.035	≤0.30	≤0.30	≤0.40	Bal.

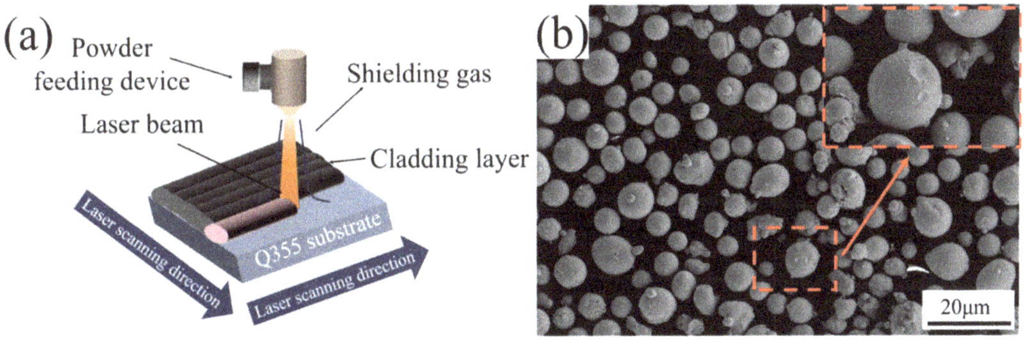

Figure 2. (**a**) Schematic illustration of the laser cladding in this study; (**b**) SEM morphologies of CoCrFeNiMo HEA powders.

As shown in Figure 3, laser deposition was executed via a LYR3063 system (LYR3063, Jinan Senfeng Technology Co., Ltd., Jinan, China) under four distinct power configurations: 1750, 2000, 2250, and 2500 W. Other process parameters were standardized as follows: 150 cm/min scanning velocity, 1.5 rpm powder feed rate, 2 mm beam diameter, and 50% overlap ratio with 1 mm/pass advancement.

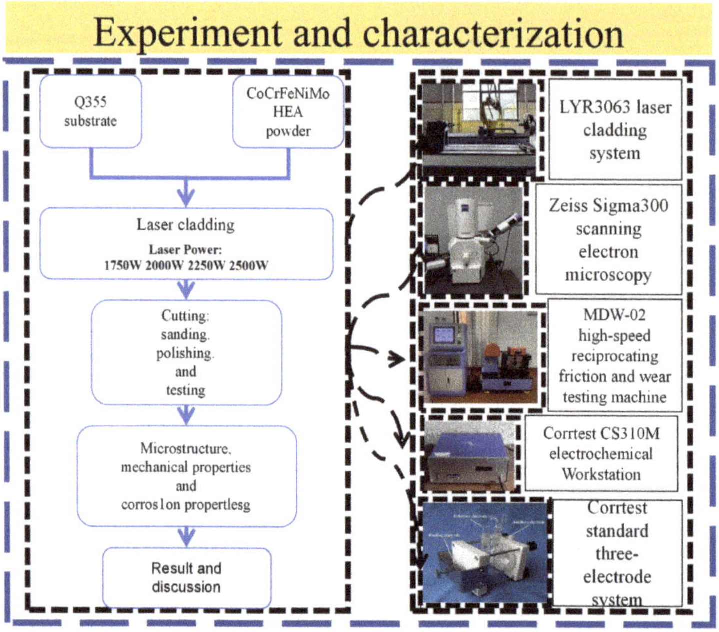

Figure 3. Experimental equipment and procedures.

2.2. Characterization of the Coatings

In this study, the coatings were first polished step-by-step. Sandpapers of grit sizes 300, 500, 1000, 1200, 1500, and 2000 were used in sequence for initial polishing. Subsequently, W1.5 diamond paste was employed to create a mirror-smooth surface on the coatings. After that, the polished samples underwent a 30 s etching process in a mixed acid solution with a HCl: HNO_3 volume ratio of 3:1.

Multi-scale characterization of the coatings was conducted employing a systematic analytical protocol: phase identification via X-ray diffraction (XRD, X'Pert PRO MPD, PANalytical B.V., Almelo, The Netherlands) with Cu Kα radiation (λ = 1.5406 Å) at 40 kV/40 mA, operating in 2θ geometry with 5°/min scanning resolution; metallographic examination using optical microscopy (OM, GX53, Olympus Corporation, Tokyo, Japan); and coupled with scanning electron microscopy (SEM, Sigma 300, Carl Zeiss AG, Oberkochen, Germany) operated in secondary electron imaging mode and back-scattered electron imaging (15 kV accelerating voltage) integrated with energy-dispersive X-ray spectroscopy (EDS, OXFORD Xplore 30, OXFORD INSTRUMENTS, London, UK) for elemental microanalysis.

Microhardness characterization was systematically conducted using a Vickers microhardness testing system (HVS-1000, Shanghai Precision Instruments Co., Ltd., Shanghai, China) under standardized indentation protocol (200 g load, 15 s dwell time) following strict compliance with ASTM E384-17 [29]. An average of 100 points were taken within a 3 mm × 3 mm square area on the surface to create a surface hardness contour map. Cross-sectional hardness profiling was performed along the coating–substrate interface normal direction using metallographically prepared cross-sections, with automated stage positioning at 0.1 mm incremental steps from the coating surface to the substrate interface, capturing the hardness gradient evolution across the coating thickness.

Tribological performance evaluation was conducted using the reciprocating high-speed friction and wear tester (MDW-02, Jinan Yihua Tribology Testing Technology Co., Ltd., Jinan, China) under dry sliding conditions, according to standard ASTM G133-05 [30].

The test protocol employed a 6 mm reciprocating stroke with 20 N load at 10 Hz oscillation frequency (equivalent to 0.12 m/s sliding velocity), generating a total sliding distance of 216 m over a 30 min test duration. Post-test wear track characterization was performed using white-light interferometry (Sneox 090, QUESTAR, St. Miami, FL, USA) to quantify three-dimensional wear volume. Wear mechanisms were analyzed through the SEM-EDS examination of wear track cross-sections.

Electrochemical corrosion behavior was quantitatively evaluated through a sequential testing protocol per ASTM G3-14 [31]. The electrochemical analysis system (CS310M, Wuhan Corrtest Instrument Co., Ltd., Wuhan, China) was configured with a conventional three-electrode array: platinum mesh counter electrode, saturated calomel reference electrode (SCE), and working electrode (coating specimen with 0.38 mm^2 geometric surface area). Open-circuit monitoring (OCP) stabilization over 3600 s was performed until potential drift < 2 mV/min. Otentiodynamic polarization: voltage scanning from −1.0 V to +0.5 V vs. OCP at a 10 mV s^{-1} scan rate, with Tafel extrapolation for corrosion current density determination.

3. Results

3.1. Phase Composition and Microstructure

According to the XRD analysis results presented in Figure 4, the CoCrFeNiMo HEA coating exhibited a coexistence of FCC and σ phases, with the FCC phase being the dominant constituent. The FCC phase was primarily composed of Ni, Cr, Co, and Mo. The σ phase, characterized as a topologically close-packed structure, displayed two distinct compositional characteristics: one predominantly consisting of Fe and Cr, and the other mainly comprising Fe and Mo [32,33].

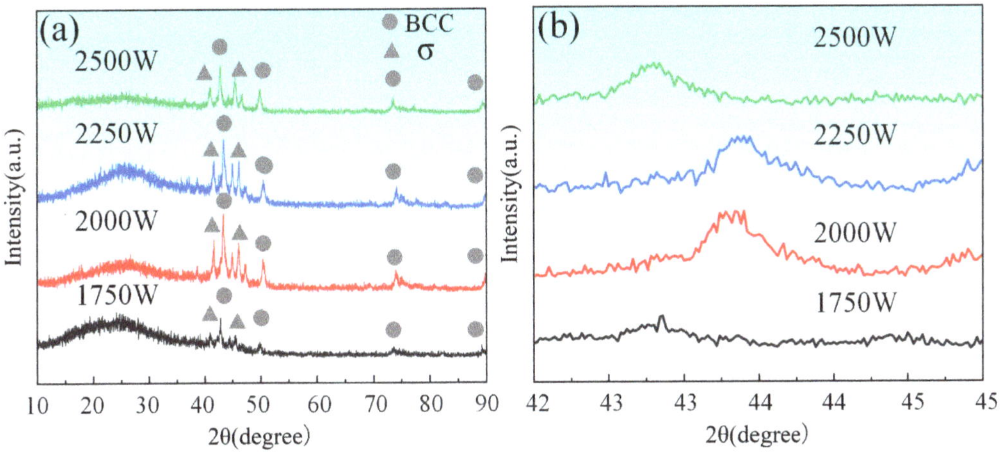

Figure 4. XRD patterns of the FeNiCoCrMo HEA coating with different laser powers: (**a**) overall view (**b**) enlarged view from a specific angle.

Furthermore, as evident from Figure 4b, the diffraction peak intensity initially increased and subsequently decreased with increasing laser power. Simultaneously, the diffraction peak positions initially shifted noticeably to the right and then to the left. These variations likely reflect alterations in the lattice parameters and dynamic adjustments in the stress state within the coating. The observed peak shifts are also correlated with changes in interatomic spacing in the coating [34].

As reported in Reference [32], for $Mo_xCoCrFeNi$ HEA, the formation of additional Mo-containing phases beyond the FCC phase was observed when the molar fraction of Mo exceeded 0.3. This phenomenon occurs because in HEA systems composed of Fe, Ni, Co, Cr, and Mo, higher Mo content tends to suppress FCC phase formation. In this study, Q355 steel was employed as the substrate, with Fe being its primary constituent. With increasing laser power, more Fe from the substrate was incorporated into the coating. This led to an elevated Fe content in the coating, which subsequently promoted the formation of Fe- and Mo-containing σ phases, thereby reducing the diffraction intensity of the FCC phase.

The FCC phase represents a typical solid solution structure, whereas the σ phase belongs to the topologically close-packed (TCP) phase category. In HEA systems, TCP phases tend to form when the electro-negativity difference (ΔX) exceeds 0.133. The ΔX can be calculated using the following Equation (1) [35]:

$$\Delta X = \sqrt{\int_{i=1}^{N} C_i(X_i - \overline{x})^2} \quad (1)$$

Through calculation, the ΔX of the CrCoFeNiMo HEA was determined to be 0.1612. Compared with the single FCC phase observed in the CrCoFeNi HEA, the CrCoFeNiMo HEA coatings exhibited a dual-phase structure consisting of FCC and σ phases.

When Mo is added to the CoCrFeNi HEA, it first dissolves into the FCC lattice. Due to Mo's large atomic radius, it causes severe lattice distortion in the FCC phase. When the solubility limit of Mo in the FCC phase is reached, Mo-rich σ phase precipitates within the FCC matrix. During the solidification of the CoCrFeNiMo HEA coating, the σ phase exhibits different morphologies depending on the temperature at which it precipitates during the eutectic reaction. Under higher laser power conditions, Mo elements were sufficiently melted and formed alloy compounds with other constituent elements, inducing lattice distortion in the FCC phase. This distortion manifested as a rightward shift in the diffraction peaks. Additionally, the increase in laser power caused a higher dilution rate of the coating, particularly due to the enhanced incorporation of Fe from the Q355 substrate into the coating, thereby further elevating the Fe content in the coating. The increased laser power also improved the fluidity of the molten pool, facilitating better mixing between different phases. However, excessively high laser power led to excessive participation of Fe in the formation of the HEA, which might cause a leftward shift in the diffraction peaks, reflecting further alterations in lattice parameters.

Figure 5 illustrates OM images of the CoCrFeNiMo HEA coating, corresponding to different laser power settings: 1750 W, 2000 W, 2250 W, and 2500 W. The images clearly revealed the presence of pores, through-thickness cracks, and interfacial cavities in the coatings produced at 1750 W, 2000 W, and 2500 W.

At the laser power of 1750 W, large pores were observed at the top surface of the coating, while fine pores and transverse cracks were present at the coating–substrate interface. Additionally, through-thickness cracks were identified, resulting in poor forming quality. When the power was increased to 2000 W, pores remained in the coating, but the number of cracks significantly decreased, with through-thickness cracks disappearing. However, interfacial cavities emerged at the coating–substrate interface. At 2250 W, coating defects were substantially reduced, with only pores being observed, demonstrating significantly improved forming quality compared to the other three parameters. However, at the highest power of 2500 W, coating defects increased again, exhibiting not only pores but also pronounced interfacial cavities.

During the laser heating and cooling process, the powder and substrate were melted by the high-energy laser beam. Upon cooling, the cladding track solidified, and the substrate formed a large molten pool. The sides and bottom of the molten pool cooled more rapidly than the central region. When the center solidified, this difference was compensated by plastic deformation in the center. Additionally, during multi-track cladding, the maximum stress increased with the deposition of adjacent tracks. The stress mainly consisted of three components: first, the volume shrinkage from liquid to solid phase during the solidification of the HEA powder; second, the microstructural stress caused by volume contraction as the solid phase cooled to room temperature; and third, the martensitic phase transformation induced by the thermal effect on the substrate surface. These combined effects led to crack formation in the coating. Farahmand et al. [36] obtained similar results through simulation, showing significant residual tensile stress concentration within the composite coating, residual compressive stress in the heat-affected zone, and varying residual stress levels between cladding tracks, with lower residual stress at earlier cladding positions. Q355 steel undergoes a martensitic transformation near 400 °C. At a laser power of 2500 W, excessive heat input intensified the volume expansion caused by this transformation. In Wang's study [37], the martensitic transformation of the substrate increased the tensile stress in the coating and was identified as a primary cause for the formation of cavities at the bottom of the coating.

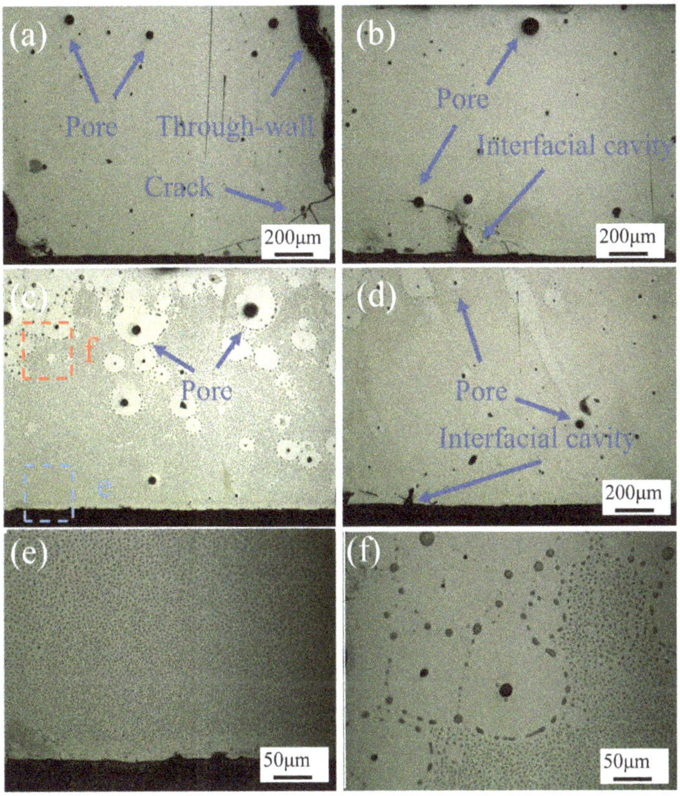

Figure 5. OM images of CoCrFeNiMo coatings with different laser powers: (**a**) 1750 W; (**b**) 2000 W; (**c**) 2250 W; (**d**) 2500 W. (**e**,**f**) present magnified views of the corresponding regions in (**c**).

At lower laser power settings, insufficient dilution between the coating and substrate prevented the formation of effective metallurgical bonding, resulting in inadequate interfacial strength. With increasing laser power, enhanced dilution between the coating and substrate promoted stronger metallurgical bonding, significantly reducing coating defects. Nevertheless, when the laser power was further increased to 2500 W, excessive heat input not only induced martensitic transformation in the substrate but also generated substantial interfacial cavities due to the mismatch in thermal expansion coefficients between the coating and substrate [38].

Figure 5e,f present magnified views of the corresponding regions in c, revealing that the CoCrFeNiMo HEA coating primarily consists of two distinct phases: a yellow phase and a black phase. In this context, the yellow phase represents the matrix, while the black phase, serving as the secondary phase, is dispersed throughout the matrix with a noticeably increasing distribution density from bottom to top. Notably, the black phase exhibits a disappearance phenomenon around the pores. Based on previous XRD analysis results, we can infer that the yellow phase corresponds to the FCC-structured matrix phase, whereas the black phase represents the σ phase, which is an enhanced phase formed through specific elemental enrichment.

Figure 6 displays the SEM images obtained under different laser power settings. Due to the poor forming quality of the coating at 1750 W, this investigation focused on three representative conditions: 2000 W (Figure 6(a_1–a_3)), 2250 W (Figure 6(b_1–b_3)), and 2500 W (Figure 6(c_1–c_3)). These images reveal a complex tri-phase microstructure comprising gray, white, and black phases. Specifically, the interfacial regions near the coating–substrate boundary (shown in Figure 6(a_1,b_1,c_1)) exhibited coarse-grained structures. The grain morphology demonstrated significant evolution with increasing laser power: gradual refinement of grain size and progressive color transition from black to gray.

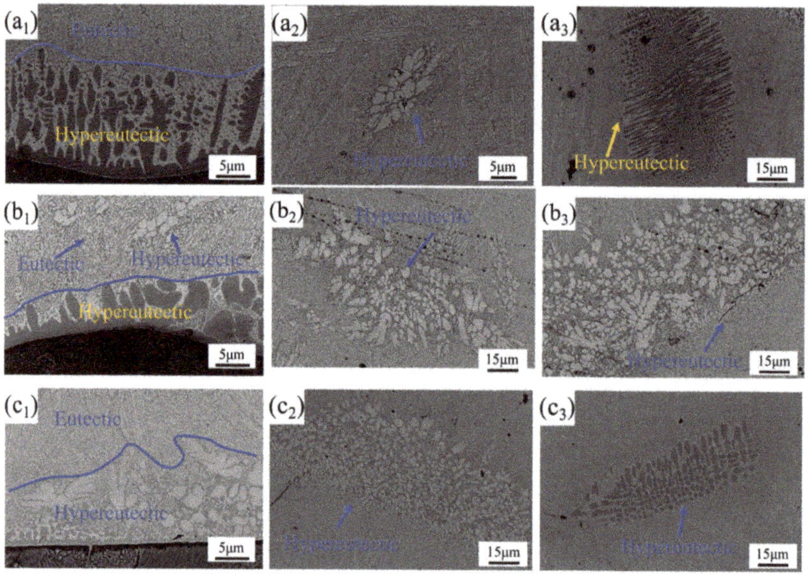

Figure 6. SEM images of CoCrFeNiMo coatings with different laser power: (**a_1–a_3**) 2000 W; (**b_1–b_3**) 2250 W; (**c_1–c_3**) 2500 W.

As shown in Figure 6(a_1,b_1,c_1), the coating's interfacial region exhibited a tri-phase (black/gray/white) microstructure comprising eutectic and hypereutectic structures. At 2000 W laser power, the near-interface zone predominantly displayed hypereutectic structures that gradually transitioned to eutectic morphology. When the power increased to 2250 W, the hypereutectic area fraction decreased, and the structure above the large grain region evolved into a eutectic–hypereutectic composite structure. Further increasing the power to 2500 W resulted in the complete disappearance of the black and a full transition from a hypereutectic structure to a eutectic structure.

The mid and upper regions of the coating (as shown in Figure 6(a_2,a_3,b_2,b_3,c_2,c_3)) exhibited a microstructure predominantly composed of eutectic structures with supplementary hypereutectic constituents arranged in a lamellar-intergranular distribution. At 2000 W, the middle and upper regions of the coating contained white lamellar reinforcing phases and black vertical/round reinforcing phases, respectively. When the power increased to 2250 W, white lamellar reinforcing phases were observed in both the middle and upper regions, with an increase in both the area fraction and grain size of these phases. As the laser power further increased to 2500 W, the area fraction of reinforcing phases in the middle region continued to expand, while their grain size decreased. Notably, black elliptical reinforcing phases reappeared in the upper region of the coating.

With the increase in laser power, the grain morphology transitioned from an initial polygonal shape to a more regular elliptical form, accompanied by a gradual enlargement of grain area and a shift in distribution from isolated dot-like patterns to continuous lamellar structures. Additionally, the originally elongated black grains tended to adopt elliptical morphologies as the power increased. These observations indicated that the thermal effects and rapid cooling under high laser power exerted a direct influence on grain morphology and distribution. The elevated laser power introduced higher thermal input, which promoted more thorough elemental fusion within the alloy and facilitated the formation of new phases. Simultaneously, it accelerated grain boundary migration rates and contributed to grain refinement.

During laser cladding, the CoCrFeNiMo HEA powder underwent rapid heating and cooling, where Mo atoms acted as solute atoms during solidification, leading to their redistribution within the FCC structure. Previous studies [25,39] demonstrated that the addition of Mo resulted in an increase in FCC lattice parameters. However, when the Mo content exceeded a critical threshold, stable FeCr-type σ phases formed in the FCC matrix. With a further increase in Mo content, FeMo-type σ phases emerged. The formation of these phases consumed a significant amount of Mo atoms, reduced the solubility of Mo in the FCC structure, and released lattice distortion energy, thereby decreasing the average lattice constant of the FCC phase. In this study, FeCr-type σ phases were referred to as σ_a phases, while FeMo-type σ phases were designated as σ_b phases.

Mo incorporation not only substantially enhanced lattice distortion but also impeded conventional atomic diffusion pathways. This dual effect effectively suppressed grain growth through two competitive mechanisms: the increased lattice distortion energy surpassed the grain boundary energy, creating an energy barrier for grain boundary migration; and the diffusion-limiting effect disrupted typical grain coarsening processes. These synergistic interactions resulted in a stabilized fine-grained microstructure. In this study, large grains formed by the σ phase were present at the bottom of the coating. As the laser power increased, the dilution rate of the coating increased, causing the σ_a phase at the bottom to transform into the σ_b phase. In the upper and middle parts of the coating, due to the different precipitation temperatures of the σ_b and σ_a phases during the solidification process, different eutectic and hypereutectic structures were formed.

A previous study [40] demonstrated that during the solidification process, the Mo content in the liquid phase exhibited a significant decrease compared to Cr, indicating the initial precipitation of the Mo-rich σ_b phase. As the temperature decreased, the FCC and σ phases co-precipitated. However, the limited content and short precipitation duration of the σ_b phase during early solidification stages hindered its growth. In the latter stage of solidification, the Cr content in the residual liquid phase decreased substantially, leading to the predominant formation of the σ_a phase. The progressive increase in σ-phase content induced the pronounced coarsening of interdendritic (ID) regions, ultimately resulting in the development of large-area network-like and blocky σ-phase configurations within the coating. In this study, the rapid solidification of the high-energy laser-induced melt pool led to the sequential precipitation and interleaved arrangement of the FCC and σ phases, forming a distinctive dual-phase architecture characterized by eutectic–hypereutectic hybrid structures.

Figure 7(a_1,b_1,c_1) present the overall morphologies of coatings prepared at 2000 W, 2250 W, and 2500 W, respectively, with a_1, b_1, and c_1 displaying their corresponding EDS line-scanning profiles and magnified views. The coating fabricated at 2250 W exhibited the fewest defects, containing only minimal pores. EDS line-scanning results revealed significant compositional fluctuations in Mo compared to other elements, further confirming the interleaved arrangement of Mo-depleted FCC phases and Mo-enriched σ phases within the coating. The magnified EDS profiles demonstrated that a distinct diffusion transition zone formed at the coating–substrate interface. The significant elemental fluctuations in this zone indicated that when the laser power was 2250 W, the diffusion zone reached its maximum length of 30 μm. At this power level, the coating demonstrated both low dilution rates and strong metallurgical bonding, effectively maintaining the designed compositional integrity of the coating. Pronounced elemental fluctuations in this region indicated robust metallurgical bonding with low dilution rates, effectively preserving the designed compositional integrity of the coating.

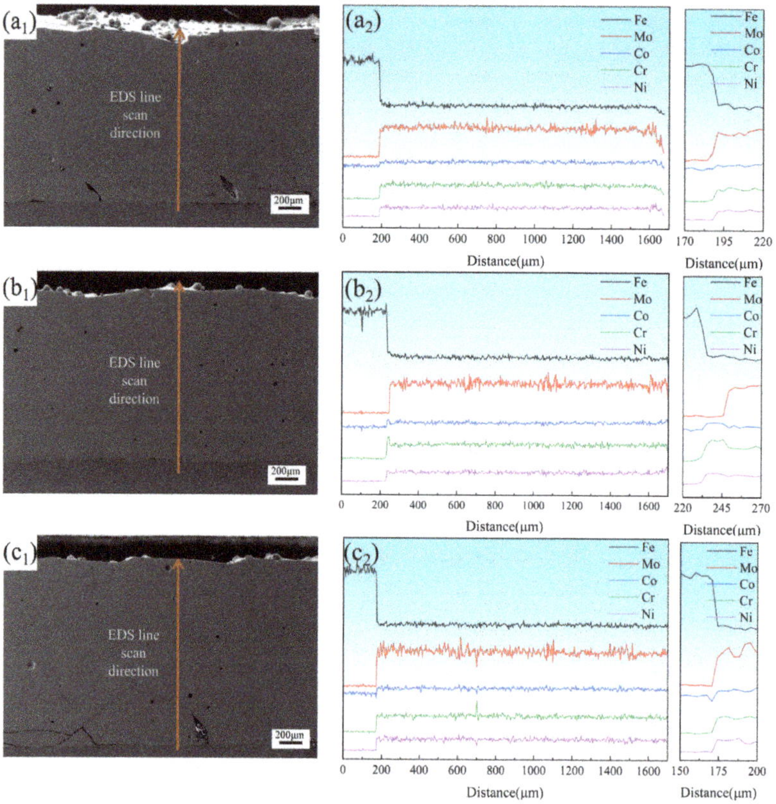

Figure 7. Overall morphology of CoCrFeNiMo HEA coatings and EDS element scans at corresponding positions: (**a_1**,**a_2**) 2000 W; (**b_1**,**b_2**) 2250 W; (**c_1**,**c_2**) 2500 W.

To further analyze the microstructural characteristics of the CoCrFeNiMo HEA coating, EDS mapping was performed on the middle region of the coating under 2250 W laser power. Figure 8 shows the results of the EDS mapping scan, where the enrichment of Mo elements in the white grains was clearly observed, while the gray region was noticeably deficient in Mo. Based on these observations, we inferred that the white grains corresponded to the σ phase, while the gray boundaries were identified as the FCC phase. The σ phase, known for its high hardness and corrosion resistance, was regarded as a strengthening phase in the alloy [41]. Simultaneously, the FCC phase exhibited excellent plasticity and toughness.

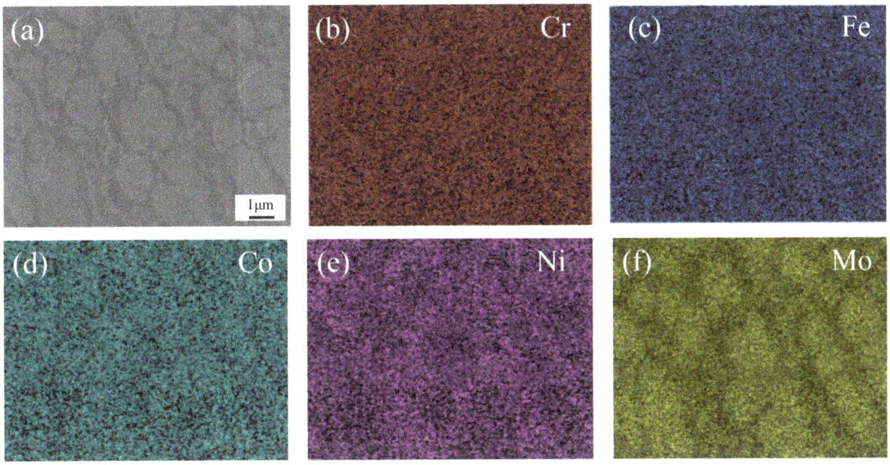

Figure 8. EDS at CoCrFeNiMo HEA coating with 1500 W power SEM morphology: (**a**) SEM morphology; (**b**) Cr; (**c**) Fe; (**d**) Co; (**e**) Ni; (**f**) Mo.

3.2. Mechanical Properties

As illustrated in Figure 9, the surface hardness analysis revealed that the overall trend of microhardness showed small variation across different conditions. As shown in Figure 9d, the Vickers hardness values measured 727.3, 733.6, and 738.3 $HV_{0.2}$ for coatings produced at 2000 W, 2250 W, and 2500 W, respectively. These results demonstrate a gradual increase in coating hardness with elevated laser power. The enhanced hardness of the CoCrFeNiMo HEA primarily originates from three strengthening mechanisms: solid solution strengthening, grain refinement strengthening, and secondary phase strengthening. Particularly, the secondary phase strengthening is predominantly associated with the σ phase, which exhibits intrinsically high hardness and effectively impedes dislocation motion within the crystal lattice. Furthermore, the eutectic regions in the alloy possess a characteristic lamellar structure that can accommodate high dislocation densities, thereby significantly contributing to the overall hardness enhancement of the alloy.

In the CoCrFeNiMo HEA coating, the microhardness distribution at 2000 W exhibited numerous low-hardness regions, indicating significant hardness inhomogeneity. When the laser power increased to 2250 W, the low-hardness regions decreased, resulting in improved hardness uniformity. However, at 2500 W, the coating demonstrated the most pronounced hardness inhomogeneity, with the maximum difference between the lowest and highest hardness values. At a power of 2000 W, due to insufficient heat input, the lifetime of the molten pool was not long enough to allow the Mo element to diffuse uniformly, resulting in a small lattice distortion effect in the FCC phase. This led to the presence of many low-hardness regions in the coating, causing hardness inhomogeneity. In contrast, at a power of 2250 W, the lattice distortion of the FCC phase was sufficient, and the distribution of the σ phase was uniform, forming a dense network-like eutectic structure, which improved the overall hardness uniformity. However, when the laser power was further increased to 2500 W, the lifetime of the molten pool increased, allowing enough time for the σ phase to precipitate during the solidification process. The extensive precipitation of the σ phase reduced the solubility of Mo in the FCC phase, resulting in a lower degree of lattice distortion and a decrease in hardness, which created a significant hardness difference throughout the entire coating.

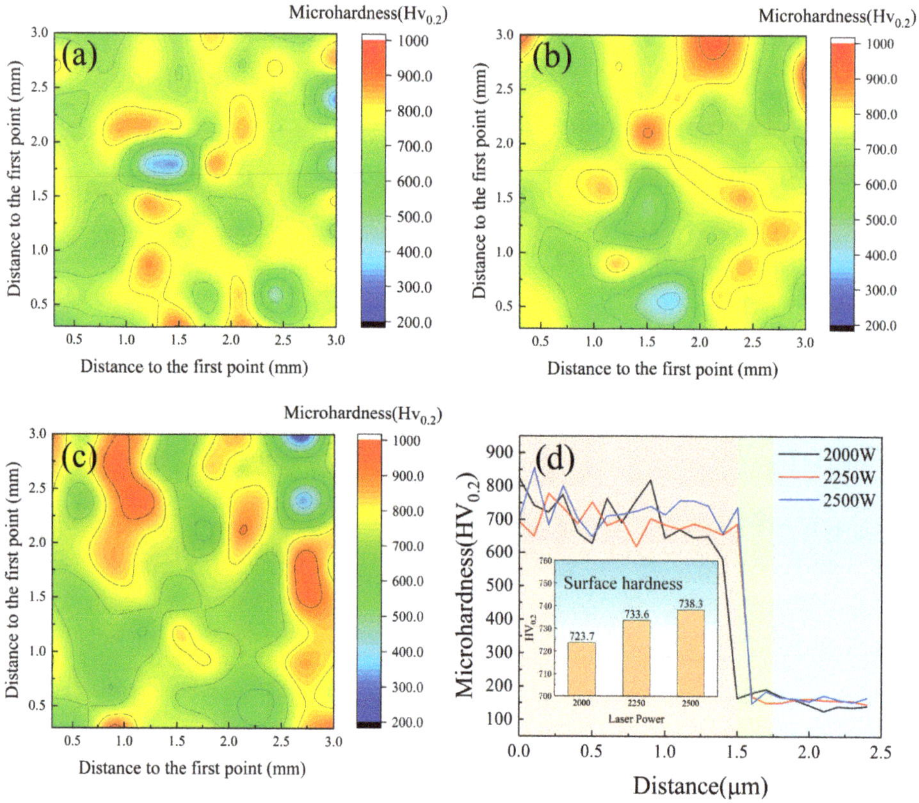

Figure 9. The hardness of CoCrFeNiMo HEA coating prepared by different laser powers: (**a**) 2000 W; (**b**) 2250 W; (**c**) 2500 W; (**d**) hardness of section.

Figure 9d demonstrates the cross-sectional hardness characteristics of the CoCrFeNiMo HEA coating under different laser power conditions, from left to right are the coating, heat-affected zone, and substrate, distinguished by different color zones, with an average coating thickness of 1.5 mm. The coating exhibited relatively uniform hardness distribution throughout the cross-section, although minor fluctuations were observed, which could be attributed to factors such as melt pool dynamics, cooling rates, and interfacial interactions with the substrate. Furthermore, the heat-affected zone displayed a comparatively limited area, indicating effective control of thermal input during the laser cladding process that minimized thermal damage and microstructural alterations in the substrate. During laser cladding, the high-energy beam and rapid non-equilibrium cooling refined the grain structure of the CoCrFeNiMo HEA coating, while σ-phase precipitation enhanced hardness to about 730 $HV_{0.2}$ because of the structure of the TCP phase. Benefiting from the intrinsic characteristics of the HEA system and precise laser processing parameters, the coating hardness reached 3–4 times that of the substrate material.

Figure 10a illustrates the schematic of the reciprocating wear test configuration, where a Si_3N_4 ceramic ball with a radius of 2 mm was employed as the sliding counterpart. The test was conducted under a normal load of 20 N at a frequency of 10 Hz for 30 min. As shown in Figure 10b, the wear volume and wear depth exhibited distinct laser power dependencies. The coatings prepared at 2000 W and 2200 W demonstrated comparable wear volumes, whereas the 2500 W sample showed significantly higher material loss.

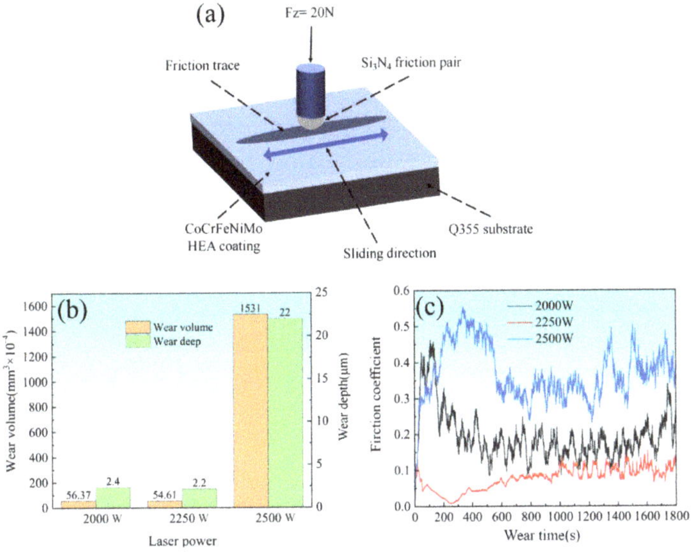

Figure 10. Schematic diagram of friction and wear experiment, friction coefficient, wear volume, and wear depth of CoCrFeNiMo HEA coatings with different laser power settings. (**a**) Schematic diagram of friction and wear experiment; (**b**) wear volume and wear depth; (**c**) coefficient of friction.

Figure 10c presents the coefficient of friction (COF) performance of CoCrFeNiMo HEA coatings prepared under different laser power conditions. The plot reveals two distinct stages during the friction and wear process: the running-in period and the steady-state wear period. During the running-in period, the COF exhibited significant fluctuations, which stabilized after approximately 200 s, marking the transition to the steady-state wear period. However, even during the steady-state wear period, the COF demonstrated notable oscillations, indicating substantial inhomogeneity in the tribological properties of the coatings. These fluctuations in friction coefficient were primarily attributed to the asperity interactions between the coating and sliding counterpart, where the micro-protrusions generated non-uniform pressure distribution across the contact surface, consequently inducing variations in frictional behavior. Furthermore, the heterogeneous distribution of phase structures (FCC and σ phases) within the coating significantly contributed to friction inhomogeneity. The FCC phase, typically demonstrating superior ductility, provided enhanced deformation accommodation, while the hard σ phase, with its characteristic high hardness, offered exceptional wear resistance through load-bearing capability enhancement.

The CoCrFeNiMo HEA coatings at different laser power settings show significant differences in friction coefficients during the steady wear stage. Specifically, the friction coefficients of the coatings at 2000 W, 2250 W, and 2500 W power settings were 0.172, 0.084, and 0.371, respectively. It is clear that the coatings at 2250 W power have the lowest friction coefficients, whereas the highest friction coefficients are found for the coatings at the 2500 W setting. These variations are mainly attributed to the differences in heat input due to different laser powers, which, in turn, affects the content and ratio of FCC, σ_a phase, and σ_b phase in the coatings. The σ phase, especially as the TCP phase, effectively resists friction and prevents crack generation and grain detachment due to its high hardness. The σ_a and σ_b phases with different size ratios and their distributions also serve as a second reinforcement. Comparatively speaking, the FCC phase, due to its higher toughness, can effectively prevent crack extension and coating delamination during frictional wear. It can be hypothesized that the insufficient heat input at a laser power of 2000 W may lead to the generation of less σ phase and, thus, a higher friction coefficient. While at 2500 W laser power, the distribution of Mo-rich σ phase in the coating was uneven, and the excessive heat input caused phase transformation in the Q355 substrate, leading to the formation of cracks at the coating interface. In contrast, at a laser power of 2250 W, the distribution and balance of the σ phase and FCC phase were better, which not only helped prevent crack formation but also effectively prevented the propagation of cracks after they formed. Therefore, by optimizing the distribution and proportion of phases, the 2250 W setting performed best in maintaining the integrity of the coating structure while providing excellent wear resistance.

Figure 11(a_1,a_2,b_1,b_2,c_1,c_2) depict the 3D morphologies and wear profiles of the coatings after tribological testing at laser power levels of 2000 W, 2250 W, and 2500 W, respectively. The wear tracks exhibited elongated geometries, with V-shaped cross-sectional profiles for the 2000 W and 2250 W coatings showing relatively consistent depth from left to right. In contrast, the 2500 W coating displayed U-shaped wear cross-sections with pronounced depth fluctuations compared to the other parameters. Wear rates were calculated using Equation (2) [42]. The wear rates of the CoCrFeNiMoHEA coatings produced at 2000, 2250, and 2500 W were 4.691×10^{-5}, 4.550×10^{-5}, and 1.275×10^{-3} $mm^3N^{-1}m^{-1}$, respectively.

$$P = \frac{V}{LS} \tag{2}$$

where P represents the wear rate ($mm^3N^{-1}m^{-1}$), V denotes the wear volume (mm^3), L stands for the applied load (N), and S signifies the total reciprocating sliding distance (m).

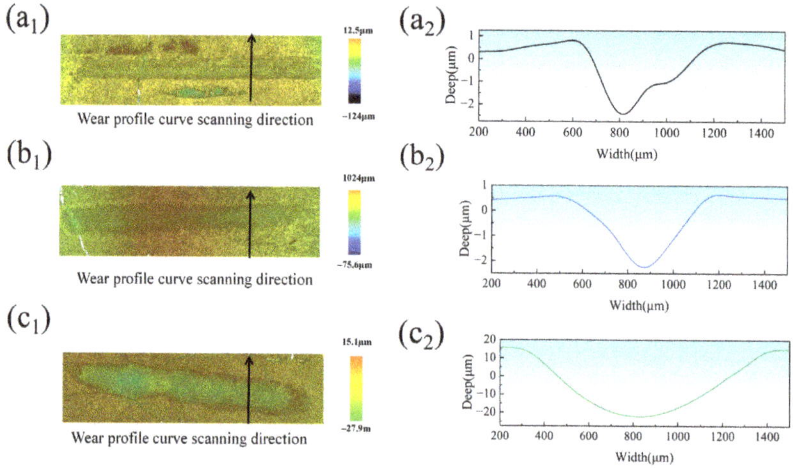

Figure 11. Three-dimensional wear morphology and wear profile of CoCrFeNiMo HEA coatings with different laser power settings: (**a_1,a_2**) 2000 W; (**b_1,b_2**) 2250 W; (**c_1,c_2**) 2500 W.

Figure 12 shows the wear morphology of the CoCrFeNiMo HEA coatings fabricated at different laser power levels after friction-wear tests. Figure 12a presents the wear morphology of the coating prepared at 2000 W laser power. In Region I, a large number of debris and adhesion marks are clearly visible, while in Region II, oxidation and grain spalling are observed. In Region III, larger grain spalling is present. EDS point scans were performed at points A and B in Region II. The results, shown in Table 2, indicate that the oxide formed on the coating is primarily Fe oxide, while Point B is rich in the Mo phase. During dry friction, a significant amount of heat was continuously generated, leading to the formation of an oxide film. The formation of the oxide film reduced the adhesion between the sliding surfaces, thereby effectively reducing the friction coefficient. As friction continued, the content of the oxide film gradually increased, resulting in the formation of adhesion marks. When the oxide film reached a certain level, the elements that form the oxide film were nearly depleted. As the oxide film cracked, it could no longer effectively protect the coating, and the hard phases began to peel off as a whole. The wear mechanism of the 2000 W coating is primarily oxidative wear.

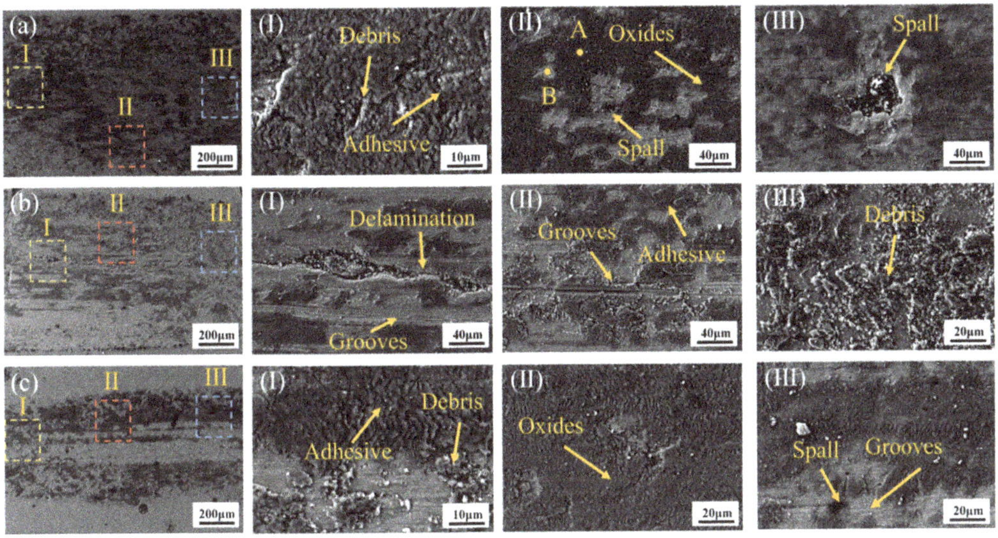

Figure 12. Wear morphologies of CoCrFeNiMo HEA coatings with different laser power settings: (**a**) 2000 W; (**b**) 2250 W; (**c**) 2500 W.

Table 2. Element content of wear morphologies (at.%).

Point	Co (at.%)	Cr (at.%)	Fe (at.%)	Ni (at.%)	Mo (at.%)	O (at.%)
A	4.88	7.09	16.39	5.67	4.63	61.34
B	18.00	18.84	18.54	17.31	18.45	8.87

Figure 12b illustrates the wear morphology of the coating prepared at 2250 W. Notably, compared to 2000 W, the content of the oxide film decreased. Grooves and delamination features were predominantly observed in Region I. This transition suggests that the 2250 W coating's elevated hard-phase fraction (σ-phase) intensified brittle spallation phenomena while enhancing abrasive wear. The fragmented oxide layers were progressively comminuted into fine debris during sliding, as evidenced by SEM mapping showing discontinuous oxygen distribution. In summary, the wear mechanisms of the coating produced at 2250 W were primarily a combination of abrasive wear, oxidative wear, and adhesive wear.

Figure 12c illustrates the wear morphology of the coating prepared at 2500 W. As observed in Figure 12c, the formed oxides were predominantly concentrated along the flanks of the wear track, while the central region exhibited extensive grooving and spallation features. In the 2500 W coating, the oxide films failed to provide adequate protection, as evidenced by their discontinuous morphology and reduced adhesion strength. The detachment of hard σ-phase particles during sliding created abrasives that exacerbated subsurface cracking, resulting in significant material loss. The dominant wear mechanism transitioned to abrasive wear supplemented by localized oxidative wear.

To further investigate the friction and wear mechanism of CoCrFeNiMo HEA coatings, an EDS mapping was conducted on the oxide layer formed after the coating's friction and wear test. From Figure 13a, it is clearly observed that the black areas are oxide formations, primarily iron oxides, which some studies [43,44] suggest are γ-Fe_2O_3. The gray areas are enriched with Mo elements, which are presumed to be the σ-phase. During the process of resisting friction and wear, the ductile FCC phase both stabilizes the hard σ-phase to prevent large-scale grain loss and forms an oxide layer that reduces the wear rate. Meanwhile, the σ-phase, due to its extremely high hardness, limits the generation and propagation of cracks in the FCC phase.

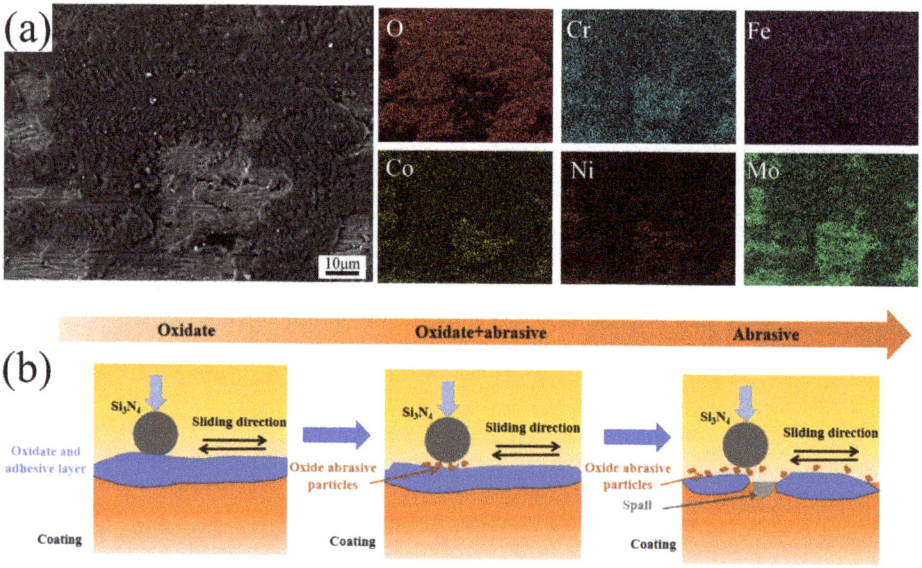

Figure 13. Wear morphologies and wear principle diagram of CoCrFeNMo coatings. (**a**) Wear morphologies; (**b**) wear principle diagram.

Figure 13b presents a schematic diagram of the CoCrFeNiMo coating's wear mechanism. During the sliding wear process, the high-speed friction between the Si_3N_4 ceramic ball and the HEA coating generated substantial frictional heating, gradually elevating the contact temperature. This thermal activation promoted the reaction between atmospheric oxygen and the elements in the HEA coating, resulting in the formation of a protective oxide layer. Wang et al. [45], through Raman spectroscopy, found that the oxides were mainly composed of Fe_3O_4 and Fe_2O_3, with the presence of MoO_3, NiO, Cr_2O_3, and Co_3O_4 as well. Among these, Fe_3O_4 and Fe_2O_3 played a positive role in the tribological properties of the HEA coating by reducing the coefficient of friction and enhancing frictional stability. However, prolonged testing induced progressive stress accumulation within the oxide layer, triggering adhesive wear at the tribo-interface. As the experiment advanced, cyclic shear stresses fragmented the oxide film into abrasive debris. Deprived of oxide protection, the hard σ-phase particles underwent brittle spallation during subsequent sliding, generating micron-scale pits that accelerated material loss.

3.3. Corrosion Properties

Figure 14 shows the results of open circuit potential (OCP) and polarization curve tests performed in 3.5 wt.% NaCl solution, which evaluated the corrosion behavior of CoCrFeNiMo HEA coatings prepared with different laser powers. As depicted in Figure 14a, the CoCrFeNiMo coatings with 2000 W showed a significant potential change process in the OCP tests, specifically a decrease followed by an increase and then a decrease again, which reflected the dynamic change of the coating surface passivation film; the initially formed passivation film may be locally dissolved under the action of corrosion medium and subsequently re-formed under electrochemical action and eventually destroyed again. In contrast, the CoCrFeNiMo HEA coatings of 2250 W and 2500 W show a relatively stable and slightly decreasing trend during the OCP test, which indicates that the passivation film formed on the surface of the coatings is able to provide a more sustained protective effect throughout the test period. In terms of specific values of corrosion potential, the CoCrFeNiMo coating at the 2000 W setting demonstrated the highest corrosion potential, indicating the lowest corrosion tendency.

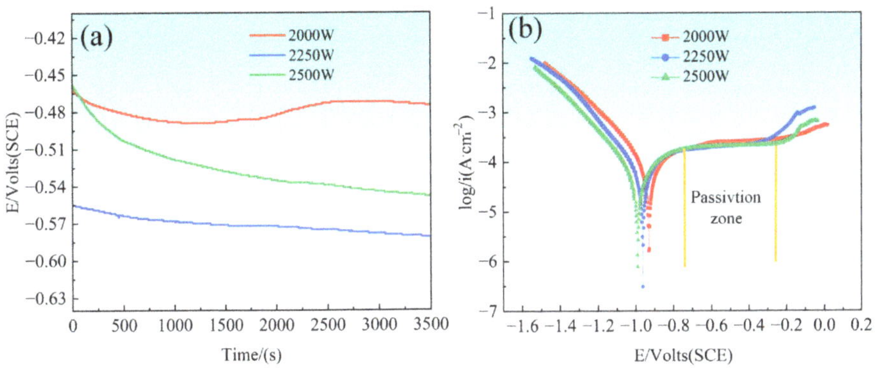

Figure 14. OCP and Tafel polarization curves of CoCrFeNiMo HEA coatings with different laser powers in 3.5 wt.% NaCl solution. (**a**) OCP; (**b**) Tafel polarization curves.

The polarization curves shown in Figure 14b demonstrate the corrosion behavior of CoCrFeNiMo HEA coatings prepared at different laser powers in 3.5 wt.% NaCl solution. These curves reveal extremely stable passivation zones exhibited by the coatings at each power setting, and this stability indicates the formation of a durable passivation film on the surface of the coatings during polarization. The CoCrFeNiMo coatings show better stability during passivation, i.e., the current density remains almost constant during the increase in corrosion potential. In Sha's study [46], Mo addition effectively refined the grain structure of HEAs and enhanced elemental distribution homogeneity, thereby reducing localized electrochemical potential differences.

The observation of pitting potentials reveals that the CoCrFeNiMo coating prepared at 2250 W laser power first reaches its pitting potential, indicating the earliest initiation of localized corrosion processes under this condition. In contrast, the 2500 W coating subsequently attains its pitting potential, while the 2000 W sample demonstrates the highest pitting potential, signifying its delayed onset of pitting corrosion at higher potentials and, consequently, superior resistance to localized corrosion in this specific test. The laser power not only governs the microstructure and phase composition of CoCrFeNiMo coatings but also significantly modulates their corrosion resistance, particularly in terms of passivation behavior and pitting resistance.

Table 3 presents the corrosion parameters of CoCrFeNiMo HEA coatings determined by the Tafel extrapolation method, including corrosion potential (E_{corr}) and corrosion current density (I_{corr}) under different laser power settings. The results demonstrate that both E_{corr} and I_{corr} values across the three laser power conditions exhibit minimal variation. Notably, the 2250 W coating displays the highest E_{corr} (−0.926 V vs. SCE) and the lowest I_{corr} (2.31×10^{-4} A·cm^{-2}), confirming its optimal corrosion resistance among the tested conditions. Corrosion current density serves as a critical indicator for evaluating the corrosion rate of materials per unit time, where lower I_{corr} corresponds to reduced corrosion rates and enhanced durability. Consequently, the 2250 W parameter not only achieves superior corrosion potential but also demonstrates a 24% reduction in corrosion rate compared to the 2500 W counterpart.

Table 3. The Tafel fitting parameters of the polarization curves.

Sample	E_{corr} (mV/SCE)	I_{corr} (mA/cm^2)
2000 W	−0.930	2.98×10^{-4}
2250 W	−0.926	2.31×10^{-4}
2500 W	−0.990	3.70×10^{-4}

To further investigate the corrosion mechanisms of the coating, two typical post-corrosion morphologies were selected, as shown in Figure 15. Figure 15a exhibits grain spallation, while Figure 15b displays a pitting corrosion pit. EDS analysis of the spalled grain in Figure 15a revealed a Cr-Fe-Co-O-enriched composition, with the surrounding matrix demonstrating Mo enrichment. As corrosion progressed, the selective dissolution of Cr, Ni, and Co occurred, forming pitting cavities as observed in Figure 15b. In the CrCoFeNiMo HEA coating, Mo acts as a chemically stable passivating element [47]. However, the electrochemical potential difference between the Mo-rich σ-phase and the Mo-depleted FCC matrix induced galvanic corrosion, accelerating localized degradation.

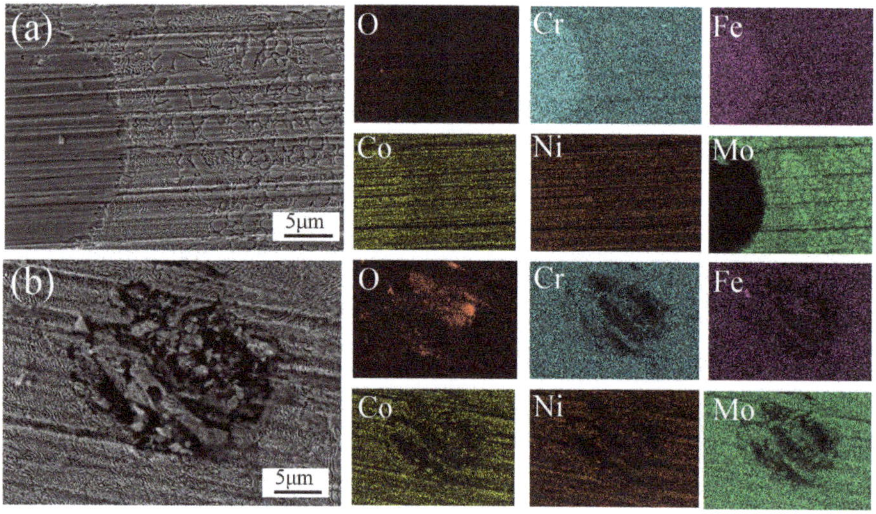

Figure 15. Typical morphologies of CoCrFeNiMo HEA after coating corrosion. (**a**) Grain spallation; (**b**) pitting corrosion pit.

Figure 16 illustrates the electrochemical corrosion progression of the CoCrFeNiMo HEA coating. The corrosion resistance of passivated alloys is intrinsically linked to the quality of their passive films. The passivation process generally proceeds through three sequential stages: 1. Anodic dissolution: The base alloy material dissolved via anodic reactions, releasing metal cations into the near-surface equiconcentration layer. 2. Cation diffusion: These cations subsequently diffused outward from the alloy surface, driven by concentration gradients and electrochemical potential. 3. Oxide nucleation and growth: Insoluble oxides/hydroxides (e.g., Cr_2O_3, MoO_3) nucleated and coalesced on the surface, ultimately forming a continuous protective passive film.

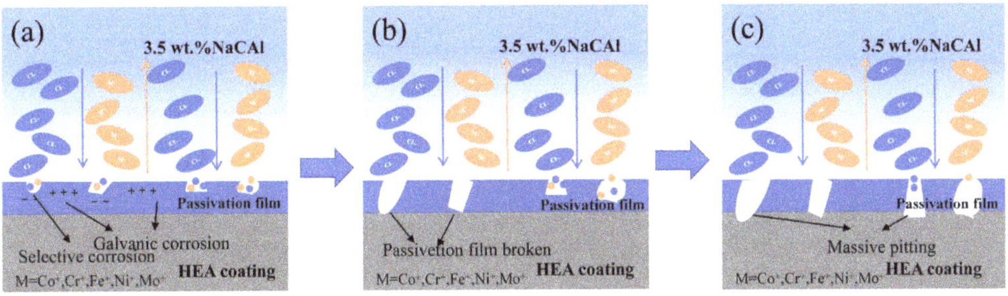

Figure 16. Electrochemical corrosion progression of the CoCrFeNiMo HEA coating. (**a**) First stage of corrosion; (**b**) second stage of corrosion; (**c**) third stage of corrosion.

In the CoCrFeNiMo HEA coating, the potential difference between the Mo-depleted FCC phase and the Mo-enriched σ phase created localized galvanic cells. This potential gradient accelerated the oxidation of Co/Fe in the FCC phase region, and the oxide film formed by Co/Fe could not effectively protect the coating, dissolving during the corrosion process. In addition, this potential difference facilitated the transfer of electrons at the phase boundaries, accelerating the selective dissolution of the coating.

The presence of Mo and Cr elements in the alloy, particularly their role in passive film formation, was of significant importance [48]. Cr was especially recognized as a crucial component for corrosion resistance due to its ability to form a stable Cr_2O_3 oxide layer. This oxide layer effectively prevented direct contact between corrosive media and the alloy matrix, thereby enhancing the overall corrosion resistance. However, the passive film underwent localized breakdown under pitting attack, exposing the underlying material and compromising its protective capability. The high corrosion potential facilitated the anodic dissolution of the matrix, potentially leading to the formation of metallic cations such as Mo^{3+}. Although the hydrolysis of these cations could have induced severe acidification within the pit cavity, molybdenum oxides and their salt compounds formed and deposited within the pits, effectively inhibiting further pit propagation. Through this mechanism, the CoCrFeNiMo HEA coating demonstrated superior corrosion resistance in aggressive environments.

4. Conclusions

This study systematically investigates the microstructure evolution, tribological properties, and corrosion mechanisms of CoCrFeNiMo high-entropy alloy (HEA) coatings fabricated via coaxial powder-fed laser cladding under variable laser power conditions. The key findings and implications are summarized as follows:

(1) The CoCrFeNiMo coatings exhibited a dual-phase structure comprising FCC and σ phases. At 2250 W, optimized energy density promoted homogeneous dispersion of Fe-Cr-rich σ_a and Fe-Mo-rich σ_b phases within the FCC matrix, forming a refined eutectic–hypereutectic hybrid microstructure.

(2) The hardness of the CoCrFeNiMo HEA coatings increased with laser power, peaking at 738.3 $HV_{0.2}$ at 2250 W. This improvement was attributed to solution strengthening and second-phase intensification. However, at 2000 W, the hardness decreased due to grain coarsening caused by the extended molten pool duration.
(3) At a laser power of 2250 W, the CoCrFeNiMo HEA coating exhibited the lowest coefficient of friction (COF = 0.084) and excellent wear resistance. This can be attributed to the oxide film formed during friction and wear, along with the eutectic structure composed of the FCC and σ phases, with the wear mechanism being oxidative wear.
(4) Electrochemical tests revealed that the CoCrFeNiMo HEA coatings achieved the best passivation film properties when at a laser power of 2250 W. The lowest corrosion current density (I_{corr} = 2.31 × 10^{-4} A/cm^2) and the highest pitting potential are attributed to the formation of the passive film of Cr/Mo, as well as the reduction in the potential difference between the FCC (Mo-poor) and σ (Mo-rich) phases.

The CoCrFeNiMo HEA coating produced at 2250 W exhibits excellent mechanical and electrochemical performance, indicating strong potential for industrial application. The coaxial powder-fed laser cladding technique used in this work is compatible with industrial-scale processing due to its scalability, automation potential, and adaptability to complex geometries. Combined with the enhanced wear and corrosion resistance of the coating, this technique offers a promising solution for extending the service life of components in demanding environments, such as the marine or aerospace industries.

Author Contributions: S.L., J.D. and F.N.: performed the experiments and wrote the manuscript; H.D.: contributed to the conception and design of the study; G.M. and Y.G.: helped perform the analysis with constructive discussions; T.C. and C.X.: played an important role in interpreting the results. All authors have read and agreed to the published version of the manuscript.

Funding: This work was financially supported by the fund of the State Key Laboratory of Clean and Efficient Turbomachinery Power Equipment (No. DEC8300CG20221028EE280287) and the fund of the High-end Foreign Experts Introduction Project of Henan Province (HNGD2025027).

Institutional Review Board Statement: Not applicable.

Informed Consent Statement: Not applicable.

Data Availability Statement: The original contributions presented in this study are included in the article. Further inquiries can be directed to the corresponding authors.

Conflicts of Interest: Authors Fuheng Nie, Guijun Mao, Yang Guo, Tianlan Cao and Chong Xiang were employed by the company Dongfang Electric Corporation Dongfang Turbine Co., Ltd. The remaining authors declare that the research was conducted in the absence of any commercial or financial relationships that could be construed as a potential conflict of interest.

References

1. Li, S.; Liu, X.; Hou, X.; Liu, Z.; Wang, X.; Narayanan, J.A.; Wu, T.; Bai, Y.; Dong, Y.; Jiang, H. Evolution of corrosion mechanism of 3d transition metal high entropy alloys: A review. *J. Mater. Res. Technol.* **2025**, *35*, 4142–4163. [CrossRef]
2. Sun, Z.; Tan, X.; Wang, C.; Descoins, M.; Mangelinck, D.; Tor, S.B.; Jägle, E.A.; Zaefferer, S.; Raabe, D. Reducing hot tearing by grain boundary segregation engineering in additive manufacturing: Example of an Al_xCoCrFeNi high-entropy alloy. *Acta Mater.* **2021**, *204*, 116505. [CrossRef]
3. Li, S.; Sun, Y.; Liu, Z.; Zhou, L.; Wu, T.; Bai, Y. Evolution of tribological properties of laser cladding 3D transition metal high entropy alloy coatings. *J. Mater. Sci.* **2025**, *60*, 4118–4147. [CrossRef]
4. Averichev, O.; Ershov, A.; Galyshev, S. Electrochemical method of deposition of an alumina coating on carbon fiber. *Ceram. Int.* **2024**, *50*, 9563–9571. [CrossRef]
5. Zhang, Q.; Shao, L.; Li, W.; Cui, S.; Shang, L.; Wang, C.; Song, Q.; Zhang, C. Comparative study of anti-corrosion properties in different CrN/WC-Co duplex coatings processed by PVD/HVOF. *Surf. Coat. Technol.* **2024**, *483*, 130799. [CrossRef]

6. Kumar, V.; Verma, R. Effect of GNP and laser-surface texturing on HVOF sprayed WC10Co4Cr coatings for high-wear resistance. *Tribol. Int.* **2023**, *178*, 108057. [CrossRef]
7. Akisin, C.; Dovgyy, B.; Bennett, C.; Pham, M.; Venturi, F.; Hussain, T. Microstructural Study of Cold-Sprayed CoCrFeNiMn High Entropy Alloy. *J. Therm. Spray Technol.* **2023**, *32*, 2539–2562. [CrossRef]
8. Meghwal, A.; Anupam, A.; Schulz, C.; Hall, C.; Murty, B.S.; Kottada, R.S.; Vijay, R.; Munroe, P.; Berndt, C.C.; Ang, A.S.M. Tribological and corrosion performance of an atmospheric plasma sprayed $AlCoCr_{0.5}Ni$ high-entropy alloy coating. *Wear* **2022**, *506–507*, 204443. [CrossRef]
9. Meghwal, A.; Anupam, A.; Luzin, V.; Schulz, C.; Hall, C.; Murty, B.S.; Kottada, R.S.; Berndt, C.C.; Ang, A.S.M. Multiscale mechanical performance and corrosion behaviour of plasma sprayed AlCoCrFeNi high-entropy alloy coatings. *J. Alloys Compd.* **2021**, *854*, 157140. [CrossRef]
10. Wen, F.; Long, Z.; Xing, Z.; Wen, P.; Zhou, P. Effect of Mo Content on Microstructure and Fatigue Properties of Ni60 Coating Produced by Plasma Cladding. *J. Therm. Spray Technol.* **2023**, *32*, 1389–1405. [CrossRef]
11. Liu, M.; Peng, Q.Q.; Zhu, X.W.; Huang, Y.F.; Wang, H.D.; Piao, Z.Y.; Ma, G.Z.; Luo, X.P.; Qiao, Q.S. Development of an Alternative to the Plasma Cladding Process: The Plasma Spray Melting Technology. *J. Therm. Spray Technol.* **2023**, *32*, 1362–1377. [CrossRef]
12. Liu, N.; Lan, Z.; Liu, L.; Dou, J.; Guo, P.; Wang, X. Optimizing 304 Stainless SteelSurface Performancewith CoCrFeMoNi High-Entropy Alloy Coating via Gas Tungsten Arc Cladding. *J. Mater. Eng. Perform.* **2023**, *32*, 9114–9120. [CrossRef]
13. Li, Y.; Tang, B.; Wu, H.; Lei, M.; Li, X.; Ouyang, H.; Wang, W. Microstructure and Wear Behavior of $AlCoCrFeNiTi_{0.5}$High Entropy Alloy Coating Prepared by Electron Beam Cladding on Ti-6Al-4V Substrate. *J. Mater. Eng. Perform.* **2024**, *33*, 11240–11249. [CrossRef]
14. Qiu, X.; Zhang, Y.; Liu, G. Microstructure and properties of $Al_2CrFeCo_xCuNiTi$ high-entropy alloy coating prepared by laser cladding. *Fenmo Yejin Cailiao Kexue Yu Gongcheng/Mater. Sci. Eng. Powder Metall.* **2013**, *18*, 735–740.
15. Wang, M.; Li, Y.; Jin, T.; Fu, H. Wear resistance optimized by heat treatment of anin-situTiC strengthened AlCoCrFeNi laser cladding coating. *Mater. Test.* **2024**, *66*, 1168–1182. [CrossRef]
16. Wang, S.; Cui, X.; Jin, G.; Liu, Y.; Wen, X.; Zheng, W.; Wan, S. Study on microstructure and wear resistance of laser cladding TiAlZrVNiX high entropy alloy coating based on Laves phase modulation. *Surf. Coat. Technol.* **2024**, *490*, 131197. [CrossRef]
17. Arif, Z.; Khalid, M.; Rehman, E.; Ullah, S.; Tariq, A. A review on laser cladding of high-entropy alloys, their recent trends and potential applications. *J. Manuf. Process.* **2021**, *68*, 225–273. [CrossRef]
18. Zhang, S.; Han, B.; Li, M.; Zhang, Q.; Hu, C.; Niu, S.; Li, Z.; Wang, Y. Investigation on solid particles erosion resistance of laser cladded CoCrFeNiTi high entropy alloy coating. *Intermetallics* **2021**, *131*, 107111. [CrossRef]
19. Zeng, X.; Liu, Z.; Wu, G.; Tong, X.; Xiong, Y.; Cheng, X.; Wang, X.; Yamaguchi, T. Microstructure and high-temperature properties of laser cladded $AlCoCrFeNiTi_{0.5}$ high-entropy coating on Ti 6Al-4V alloy. *Surf. Coat. Technol.* **2021**, *418*, 127243. [CrossRef]
20. Zhang, H.; Pan, Y.; He, Y.-Z. Synthesis and characterization of FeCoNiCrCu high-entropy alloy coating by laser cladding. *Mater. Des.* **2011**, *32*, 1910–1915. [CrossRef]
21. Zhang, B.; Zhang, Y.; Guo, S.M. A thermodynamic study of corrosion behaviors for CoCrFeNi-based high-entropy alloys. *J. Mater. Sci.* **2018**, *53*, 14729–14738. [CrossRef]
22. Li, S.; Ding, J.; Liu, Z.; Wang, X.; He, P.; Narayanan, J.A.; Du, S.; Zhou, L. Evolution of microstructure and properties of AlCoCrFeNi high-entropy alloy coatings fabricated at different laser power levels. *Mater. Today Commun.* **2025**, *46*, 112644. [CrossRef]
23. Wang, Z.; Jin, J.; Zhang, G.-H.; Fan, X.-H.; Zhang, L. Effect of temperature on the passive film structure and corrosion performance of CoCrFeMoNi high-entropy alloy. *Corros. Sci.* **2022**, *208*, 110661. [CrossRef]
24. Dai, C.; Zhao, T.; Du, C.; Liu, Z.; Zhang, D. Effect of molybdenum content on the microstructure and corrosion behavior of FeCoCrNiMox high-entropy alloys. *J. Mater. Sci. Technol.* **2020**, *46*, 64–73. [CrossRef]
25. Chang, Q.; Liu, Y.; Xia, Y.; Liu, X.; Xie, Y.; Hu, J.; Li, W. Effect of Mo on microstructure and sliding wear performance of $CoCrFeNiMo_x$ high entropy alloy coatings by laser cladding. *Mater. Res. Express* **2023**, *10*, 126514. [CrossRef]
26. Zeng, Y.; Guo, Y.; Yang, Q.; Li, Z.; Wang, Y.; Wang, G.; Guo, Z.; Chen, Z. Effect of Ti, Mn and Mo on the microstructure and properties of CoCrFeNi high entropy alloy coatings prepared by laser cladding. *Mater. Today Commun.* **2025**, *41*, 110509. [CrossRef]
27. Dai, C.; Fu, Y.; Pan, Y.; Yin, Y.; Du, C.; Liu, Z. Microstructure and mechanical properties of $FeCoCrNiMo_{0.1}$ high-entropy alloy with various annealing treatments. *Mater. Charact.* **2021**, *179*, 111313. [CrossRef]
28. *ASTM A992/A992M-11*; Standard Specification for Structural Steel Shapes. ASTM: West Conshohocken, PA, USA, 2015.
29. *ASTM E384-17*; Standard Test Method for Microindentation Hardness of Materials. ASTM: West Conshohocken, PA, USA, 2017.
30. *ASTM G133-05*; Standard Test Method for Linearly Reciprocating Ball-on-Flat Sliding Wear. ASTM: West Conshohocken, PA, USA, 2016.
31. *ASTM G3-14*; Standard Practice for Conventions Applicable to Electrochemical Measurements in Corrosion Testing. ASTM: West Conshohocken, PA, USA, 2019.

32. Ni, J.; Wen, M.; Jayalakshmi, S.; Geng, Y.; Chen, X. Investigation on microstructure, wear and friction properties of CoCrFeNiMox high-entropy alloy coatings deposited by powder plasma arc cladding. *Mater. Today Commun.* **2024**, *39*, 108807. [CrossRef]
33. Li, S.; Ai, S.; Liu, Z.; Wang, X.; Zhou, L.; Wu, T.; Bai, Y. Advancements in particle-reinforced high-entropy alloy coatings: Microstructure, mechanical properties and mechanisms. *J. Alloys Compd.* **2025**, *1029*, 180787. [CrossRef]
34. Moazzen, P.; Toroghinejad, M.R.; Cavaliere, P. Effect of Iron content on the microstructure evolution, mechanical properties and wear resistance of FeXCoCrNi high-entropy alloy system produced via MA-SPS. *J. Alloys Compd.* **2021**, *870*, 159410. [CrossRef]
35. Dong, Y.; Lu, Y.; Jiang, L.; Wang, T.; Li, T. Effects of electro-negativity on the stability of topologically close-packed phase in high entropy alloys. *Intermetallics* **2014**, *52*, 105–109. [CrossRef]
36. Farahmand, P.; Kovacevic, R. An experimental–numerical investigation of heat distribution and stress field in single- and multi-track laser cladding by a high-power direct diode laser. *Opt. Laser Technol.* **2014**, *63*, 154–168. [CrossRef]
37. Wang, D.; Hu, Q.; Zeng, X. Residual Stress and Cracking Behaviors of $Cr_{13}Ni_5Si_2$ Based Composite Coatings Prepared by Laser-Induction Hybrid Cladding. *Surf. Coat. Technol.* **2015**, *274*, 51–59. [CrossRef]
38. Sun, W. A Numerical Simulation Study on the Tensile Properties of Welds in Laser-Arc Hybrid Welding of Q355 Medium-Thick Plates. *Coatings* **2025**, *15*, 252.
39. Lou, Y.; Dai, C.; Chang, W.; Qian, H.; Huang, L.; Du, C.; Zhang, D. Microbiologically influenced corrosion of $FeCoCrNiMo_{0.1}$ high-entropy alloys by marine *Pseudomonas aeruginosa*. *Corros. Sci.* **2020**, *165*, 108390. [CrossRef]
40. Li, Y.; Zhang, J.; Huang, X.; Liu, J.; Deng, L.; Han, P. Influence of laser power on microstructure evolution and properties of laser cladded FeNiCoCrMo HEA coatings. *Mater. Today Commun.* **2023**, *35*, 105615. [CrossRef]
41. Fan, R.; Wang, L.; Zhao, L.; Wang, L.; Zhao, S.; Zhang, Y.; Cui, B. Synergistic effect of Nb and Mo alloying on the microstructure and mechanical properties of CoCrFeNi high entropy alloy. *Mater. Sci. Eng. A* **2022**, *829*, 142153. [CrossRef]
42. Zhu, R.; Zhang, P.; Yu, Z.; Yan, H.; Li, S.; Wu, D.; Shi, H.; Tian, Y. Microstructure and wide temperature range self-lubricating properties of laser cladding $NiCrAlY/Ag_2O/Ta_2O_5$ composite coating. *Surf. Coat. Technol.* **2020**, *383*, 125248. [CrossRef]
43. Ren, X.; Sun, W.; Tian, S.; Zhu, C.; Qin, M.; Yang, Y.; Wu, W. Tribological and electrochemical behaviors of FeCoNiCrMox HEA coatings prepared by internal laser cladding on 316L steel tube. *Mater. Charact.* **2024**, *211*, 113906. [CrossRef]
44. Liu, J.; Liu, H.; Chen, P.; Hao, J. Microstructural characterization and corrosion behaviour of AlCoCrFeNiTix high-entropy alloy coatings fabricated by laser cladding. *Surf. Coat. Technol.* **2019**, *361*, 63–71. [CrossRef]
45. Wang, J.; Han, B.; Chen, Z.; Hu, C.; Zhang, Q.; Wang, C. Study on tribological oxide mechanism of CoCrFeNiMo high entropy alloy. *Wear* **2023**, *526–527*, 204907. [CrossRef]
46. Sha, M.; Wang, S.; Li, S.; Jia, C.; Huang, T.; Zhu, X.; Ai, X.; Liao, X. Effect of Mo on Microstructure and Properties of AlCoCrFeNi-Mox High Entropy Alloy Coatings Prepared by Laser Cladding. *Rare Met. Mater. Eng.* **2023**, *52*, 685–690.
47. Cui, C.; Wu, M.; Miao, X.; Zhao, Z.; Gong, Y. Microstructure and corrosion behavior of CeO_2/FeCoNiCrMo high-entropy alloy coating prepared by laser cladding. *J. Alloys Compd. Interdiscip. J. Mater. Sci. Solid-State Chem. Phys.* **2022**, *890*, 161826. [CrossRef]
48. Kai, W.; Chien, F.; Cheng, F.; Huang, R.; Kai, J.; Liu, C. The corrosion of an equimolar FeCoNiCrMn high-entropy alloy in various CO_2/CO mixed gases at 700 and 950 °C. *Corros. Sci.* **2019**, *153*, 150–161. [CrossRef]

Disclaimer/Publisher's Note: The statements, opinions and data contained in all publications are solely those of the individual author(s) and contributor(s) and not of MDPI and/or the editor(s). MDPI and/or the editor(s) disclaim responsibility for any injury to people or property resulting from any ideas, methods, instructions or products referred to in the content.

materials

MDPI

Communication

Corrosion Behavior of 347H Stainless Steel in NaCl-KCl-$MgCl_2$ Molten Salt: Vapor, Liquid, and Interface Comparison

Zhiwen Liu [1,2,†], Huigai Li [1,†], Yang Wang [2], Yanjie Peng [2], Luyan Sun [2] and Jianping Liang [2,*]

1 College of Materials Science and Engineering, Shanghai University, Shanghai 200072, China; liuzhiwen@sinap.ac.cn (Z.L.); lihuigai@shu.edu.cn (H.L.)
2 Shanghai Institute of Applied Physics, Chinese Academy of Sciences, Shanghai 201800, China; wangyang@sinap.ac.cn (Y.W.); pengyanjie@sinap.ac.cn (Y.P.); sunluyan@sinap.ac.cn (L.S.)
* Correspondence: liangjianping@sinap.ac.cn
† These authors contributed equally to this work.

Abstract

The suitability of 347H stainless steel (SS347H) for chloride salt environments is critical in selecting materials for next-generation concentrated solar power (CSP) systems. This study investigated the corrosion behavior of SS347H in a ton-scale purification system with continuously flowing chloride salt under three conditions: exposure to NaCl-KCl-$MgCl_2$ molten salt vapor, immersion in molten salt, and at the molten salt surface interface. Results revealed that corrosion was most severe in the molten salt vapor, where HCl steam facilitated Cl^- reactions with Fe and Cr in the metal, causing dissolution and forming deep corrosion pits. At the interface, liquid Mg triggered displacement reactions with Fe^{2+}/Cr^{2+} ions in the salt, depositing Fe and Cr onto the surface, which reduced corrosion intensity. Within the molten salt, Mg's purification effect minimized impurity-induced corrosion, resulting in the least damage. In all cases, the primary corrosion mechanism involves the dissolution of Fe and Cr, with the formation of minor MgO. These insights provide valuable guidance for applying 347H stainless steel in chloride salt environments.

Keywords: 347H stainless steel; molten chloride salt; corrosion behavior; concentrated solar power

1. Introduction

Next-generation concentrated solar power (CSP) technology has gained increasing attention for its high operating temperatures, improved power generation efficiency, and enhanced economic benefits [1–4]. Among potential thermal storage and heat transfer media, molten chloride salts (e.g., NaCl-KCl-$MgCl_2$) are highly promising due to their excellent thermophysical properties, including low viscosity, high specific heat capacity, high thermal stability (>800 °C), and low cost (<0.35 USD·kg^{-1}) [5–9]. However, unlike commercially used molten nitrates, molten chlorides are more corrosive to metallic structural materials at elevated temperatures [10–12]. This poses a significant challenge for the selection of materials for CSP systems. The choice of materials is critical to ensure the reliability and efficiency of the system over the long term [13].

347H stainless steel (SS347H) is a strong candidate due to its excellent high-temperature mechanical properties and relatively good corrosion resistance [13–15]. It has already been used in hot-salt piping and storage tanks in second-generation CSP systems using nitrate-based salt [16,17]. However, its performance in the aggressive chloride salt environments of Gen3 CSP systems remains insufficiently understood.

https://doi.org/10.3390/ma18143412

Molten NaCl-KCl-$MgCl_2$ salts exhibit high vapor pressures at elevated temperatures, and the decomposition of hydrated chlorides and impurities can release gaseous oxides that accelerate corrosion [18–20]. A key corrosion mechanism for stainless steels is the preferential dissolution of chromium (Cr), often leading to subsurface cavity formation [21–23]. While prior studies have largely focused on corrosion in either molten salt or its vapor phase, the interface region, where complex interactions occur under real working conditions, has been less explored.

This study aims to address this gap by systematically comparing the corrosion behavior of SS347H under three conditions: in the molten salt vapor, within the molten salt, and at the interface between them. Using industrial-scale salt production equipment, corrosion mechanisms were investigated under each condition. Results show that corrosion is most severe in the vapor region, followed by the interface, and least severe in the molten salt. These findings provide valuable insights for optimizing the application of SS347H in Gen3 CSP systems.

2. Materials and Methods

2.1. Materials

The material used in this study was SS347H, cut into sheet samples measuring $150 \times 20 \times 3$ mm^3. Prior to corrosion testing, the sample surfaces were mechanically ground, polished, and then ultrasonically cleaned. The chemical composition of SS347H is shown in Table 1.

Table 1. Chemical composition of SS347H (wt.%).

Material	C	Si	Mn	Cr	Ni	Nb	Ti	N	Fe
Composition	0.05	0.26	1.58	18.06	9.40	0.47	0.005	0.010	Bal.

The molten salt used was a ternary eutectic mixture of NaCl-KCl-$MgCl_2$, consisting of 24.50 wt.% NaCl, 20.55 wt.% KCl, and 54.95 wt.% $MgCl_2$.

2.2. Experimental Procedure

Corrosion experiments were conducted using a ton-scale continuous purification system for chloride salts (Figure 1a). This system enables continuous salt circulation and impurity removal (e.g., moisture and oxygen) via controlled drying and magnesium addition. Unlike static crucible setups, the flowing salt in this system interacts dynamically with internal piping, creating variable corrosion conditions.

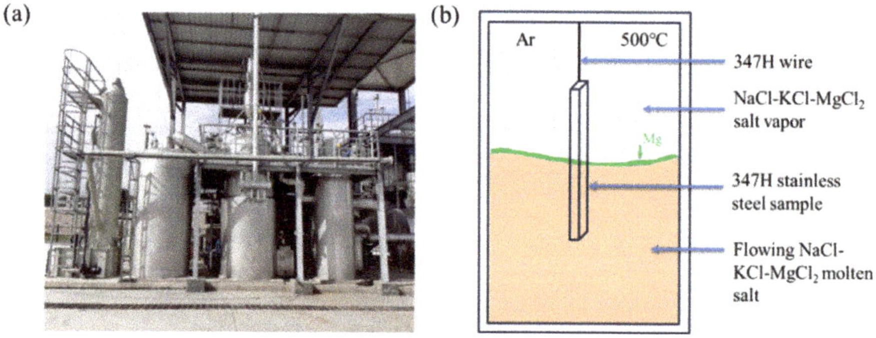

Figure 1. (**a**) Physical setup of the experimental apparatus. (**b**) Schematic of the corrosion experiment.

To simulate real-world Gen3 CSP operating conditions, the sample was suspended in the molten salt tank using SS347H wire, with one end immersed in the salt and the other exposed to argon atmosphere above the salt surface (Figure 1b). This configuration created three distinct exposure regions: (1) molten salt region (fully immersed), (2) molten salt–vapor interface, and (3) molten salt vapor region (argon + salt vapor atmosphere).

The flowing molten salt caused slight surface agitation, intermittently exposing the interface region. Inhibitory elemental magnesium, added to purify the melt, floated at the interface due to its lower density.

The corrosion test was continued for 100 h at 500 °C. After exposure, the sample was cooled and ultrasonically cleaned sequentially with tap water, deionized water, and anhydrous ethanol, and then dried with cool air. After cleaning, thickness measurements were taken at five positions within each region, with three repeated readings per position to ensure data accuracy. Corrosion behavior was evaluated separately for the vapor, interface, and molten salt regions.

2.3. Characterization

Microstructural analysis of the pristine and corroded SS347H samples was performed using X-ray diffraction (Bruker D8 Advance, Billerica, MA, USA). Morphology and elemental analysis before and after corrosion was characterized using a scanning electron microscope (SEM, Zeiss Merlin FE, Oberkochen, Germany) coupled with an energy-dispersive spectrometer (Oxford Instruments X-Max 80 mm^2, Abingdon, UK). All SEM analyses were performed at an accelerating voltage of 15 kV.

3. Results and Discussion

3.1. Macroscopic Morphology and Crystalline Structure

Upon removal from the corrosion apparatus, the surface of SS347H showed distinct visual differences across three exposure regions (Figure 2). The molten salt vapor region appeared yellow-green in appearance, the interface region featured a thick, uneven, black deposit layer, and the molten salt region had gray deposits. Macroscopic observations indicated that the most severe corrosion product accumulation occurred at the interface region.

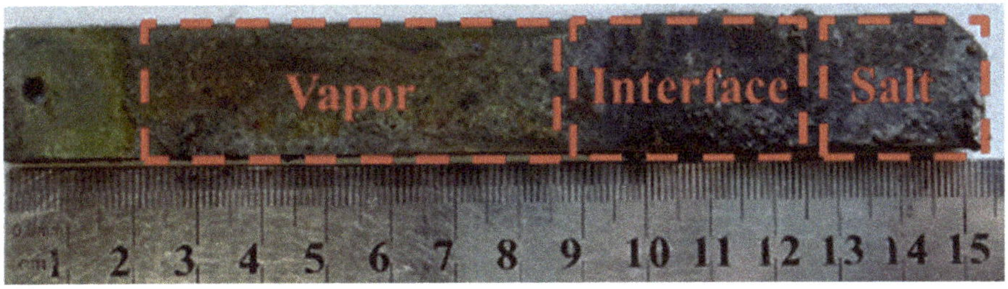

Figure 2. Macroscopic surface morphology of SS347H after corrosion in different regions.

Sample thicknesses were measured separately in each region. After corrosion and subsequent cleaning, the measured thicknesses of the metals—including the effects of corrosion products—were 3.21 mm in the vapor region, 4.39 mm at the interface, and 3.32 mm in the salt region. Given the original thickness of 3 mm, corrosion products accumulated in all regions, with the highest accumulation at the interface.

XRD analysis revealed distinct phase transformations in the three regions (Figure 3). The austenitic structure of SS347H remained predominantly stable across all regions, with additional diffraction peaks corresponding to Fe and MgO. Notably, the interface region exhibited strong Fe peaks surpassing those of the matrix, along with pronounced MgO peaks, indicating more complicated corrosion at this interface. The appearance of strong iron peaks will be discussed in Section 4.

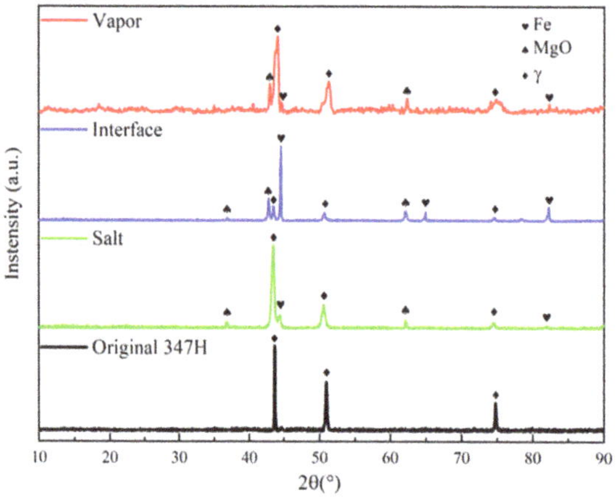

Figure 3. XRD patterns of different corrosion regions of SS347H after 100 h of exposure to NaCl-KCl-$MgCl_2$ molten salt.

3.2. Surface Morphology and Elemental Distribution

SEM images and EDS analysis (Figure 4) revealed region-specific surface characteristics and elemental distributions. In the vapor region, the surface appeared uneven, featuring irregular, deep corrosion pits (Figure 4a). EDS analysis revealed both exposed substrate (P_1) and progressive oxidation of the substrate (P_2, P_3).

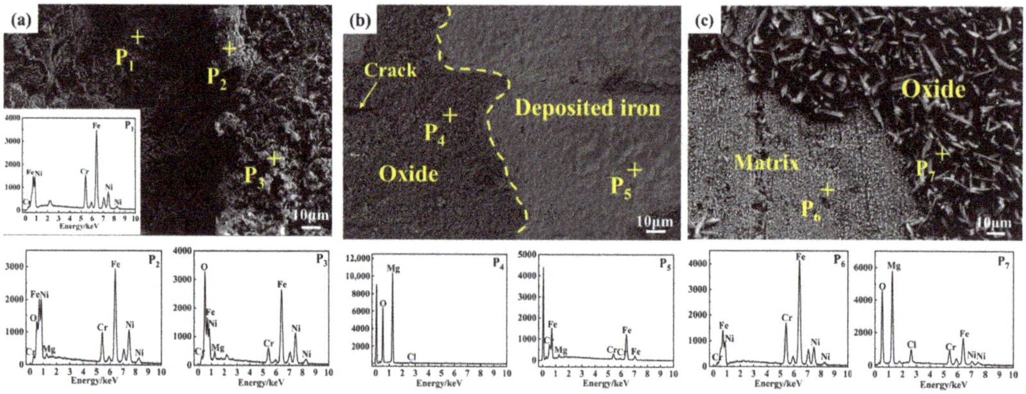

Figure 4. Surface morphology of SS347H after 100 h of corrosion in NaCl-KCl-$MgCl_2$ molten salt: (**a**) vapor region, (**b**) interface region, and (**c**) molten salt region.

In the interface region, the surface was relatively flat and exhibited two distinct areas. On the left side of Figure 4b, a dense, dark gray oxide layer showed high O and Mg content (P_4), indicating MgO formation. On the right side, a smooth, light gray area was primarily composed of Fe and Cr.

In the molten salt region, two distinct areas were also observed. Some areas showed exposed metal substrate, with Cr, Ni, Mg, and O detected by EDS (P_6). Other areas showed rough, feather-like structures, with MgO identified as the dominant component (P_7).

3.3. Cross-Sectional Morphology and Element Distribution

Cross-sectional SEM analysis (Figure 5) further clarified corrosion differences. In the vapor region, corrosion was characterized by deep, uneven corrosion pits with maximum depth 115.2 μm. In the interface region, a double-layer structure was observed: an outer granular deposit layer (pure iron and MgO) with an average thickness of 87.1 μm and a maximum of about 132.6 μm, and an inner corrosion layer of 19.8 μm. The molten salt region showed a relatively flat surface with no significant corrosion pits or defects. The total corrosion depth was 3.7 μm, consisting of a 2.6 μm outer layer and a 1.1 μm inner corrosion layer. Based on the maximum corrosion depth, the corrosion rates of SS347H after 100 h of exposure were 1.2 μm/h in the molten salt vapor region, 0.2 μm/h at the interface, and 0.04 μm/h in the salt region. The estimated corrosion rates may be higher than in real-world applications due to accelerated test conditions and the use of maximum depth. However, the aim of this study is to compare corrosion severity across the three regions, so the relative values are more meaningful than the absolute rates.

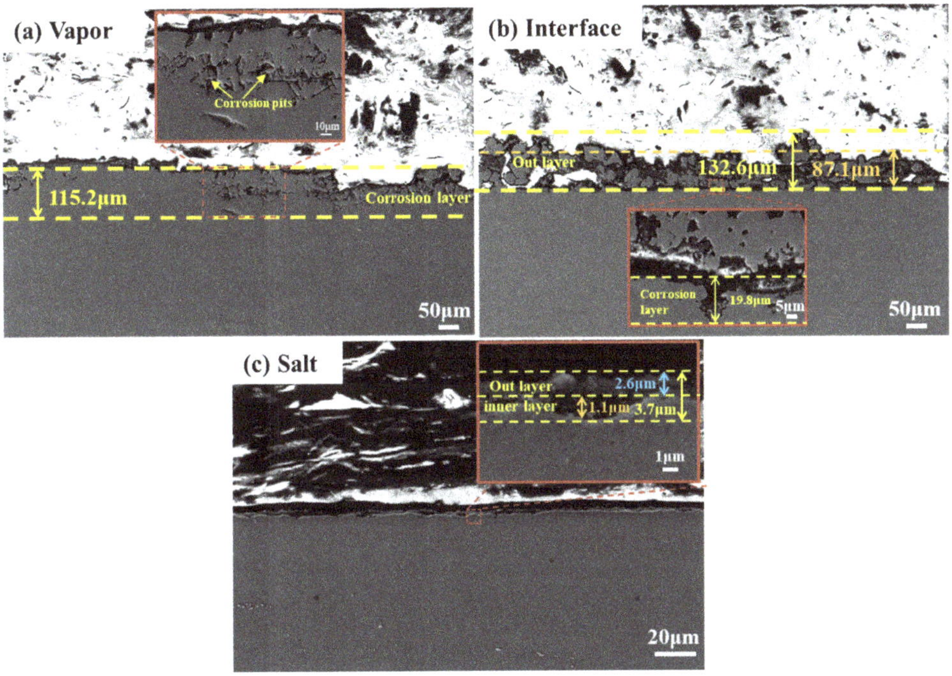

Figure 5. Cross-sectional morphology of SS347H after 100 h of NaCl-KCl-$MgCl_2$ molten salt corrosion: (**a**) vapor region, (**b**) interface region, and (**c**) molten salt region.

These results indicate that corrosion severity followed the order vapor region > interface region > salt region. This trend contrasts with the macroscopic observations discussed in Section 3.1, where the interface region appeared to exhibit the most severe corrosion. The prominent visual degradation at the interface is attributed to the accumulation of thick and uneven corrosion product layers—primarily Fe and MgO—formed through displacement reactions between Mg in the molten salt and Fe^{2+}/Cr^{2+} ions. These reactions deposit metallic Fe and Cr on the surface and promote MgO formation, leading to substantial surface buildup without significant penetration into the substrate.

In contrast, cross-sectional SEM analysis provides a direct assessment of material degradation. It reveals that the vapor region experiences the deepest corrosion penetration

(~115 μm), mainly due to the aggressive attack by volatile HCl gas on Fe and Cr. Despite its more severe surface deposits, the interface region shows a much shallower penetration depth (~20 μm), highlighting the difference between surface appearance and actual material loss.

EDS elemental mappings (Figure 6) highlighted compositional changes. In the vapor region, corrosion primarily involved the dissolution of Cr and Fe, leading to their depletion, while Ni became enriched in the corrosion pits. The line scan results (Figure 6(a_2)) confirm that Fe and Cr decreased in the pits, while Mg, O, and Ni concentrations increased. Mg and O infiltrated the corrosion region, forming MgO, with some oxide adhering to the surface. In the interface region, significant Fe deposition and MgO formation were observed (Figure 6(b_1,b_2)), along with some Cr enrichment, distinguishing it from the corrosion observed in the vapor region. The corrosion inhibitor Mg effectively reduced corrosion in the interface and salt regions but had a limited effect on vapor-phase corrosion. In the molten salt region, minimal corrosion was observed, with only a small amount of corrosion products in the outer layer and depletion of Cr and Fe in the inner layer. The outer corrosion layer had increased concentrations of Mg, O, and Cr, indicating the presence of $MgCr_2O_4$. The inner layer contained MgO, confirming the reaction of Mg and O in this region.

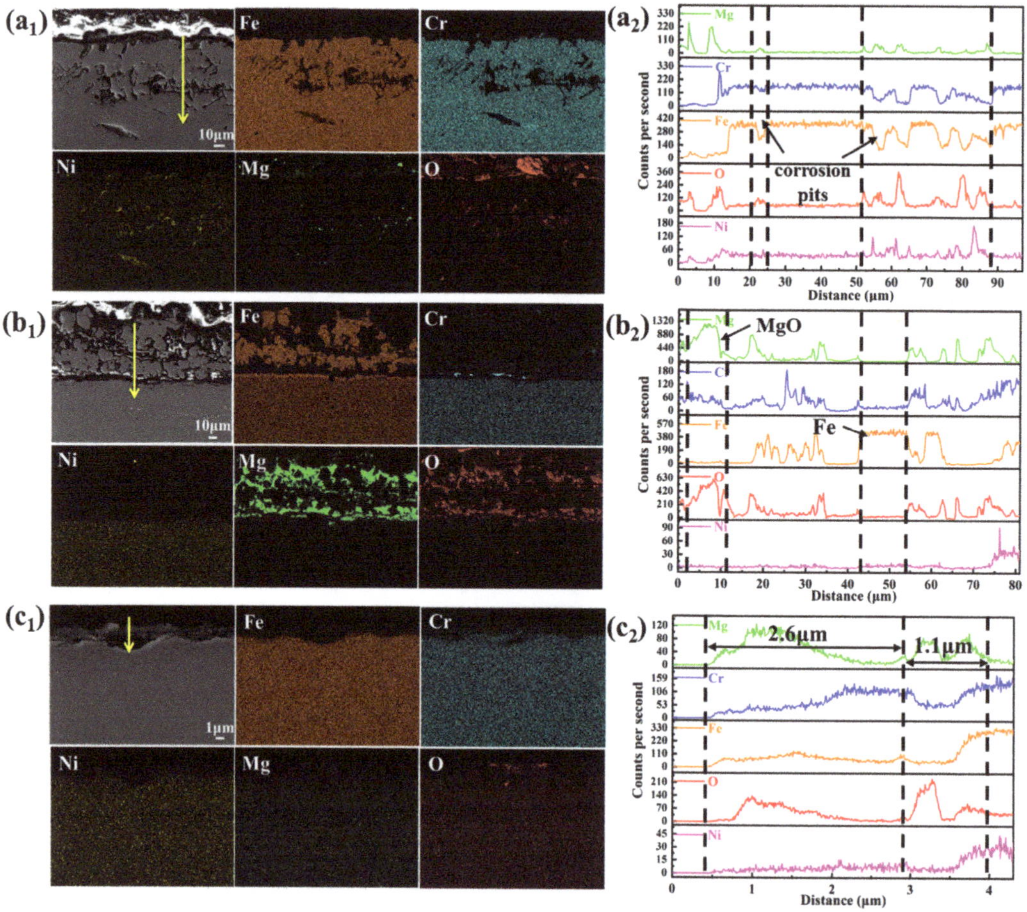

Figure 6. Element distribution and EDS results of SS347H after corrosion in three regions: (a_1,a_2) vapor, (b_1,b_2) interface, and (c_1,c_2) molten salt.

4. Corrosion Mechanism Discussion

The corrosion mechanism of SS347H in molten chloride salt is primarily governed by reactions between metallic elements (Fe and Cr) and corrosive species such as MgO-HCl, which forms predominantly due to the hygroscopic nature of $MgCl_2$. At elevated temperatures, MgOHCl decomposes into MgO and highly corrosive HCl, significantly accelerating the corrosion process. The corresponding chemical reactions are shown as Equations (1) and (2) in Figure 7.

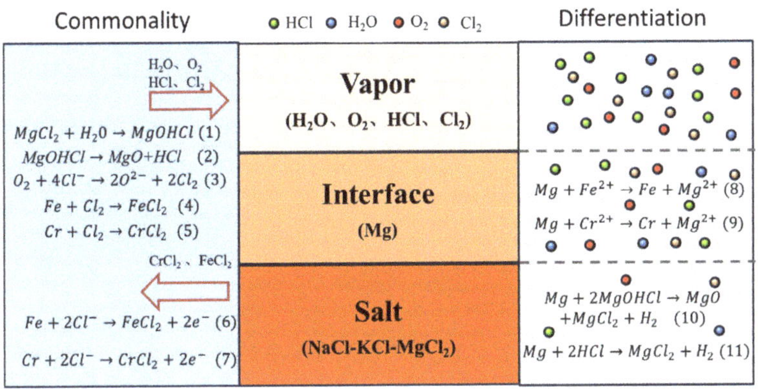

Figure 7. Schematic corrosion mechanism of SS347H in NaCl-KCl-$MgCl_2$ molten salt.

The generated HCl volatilized into the vapor phase, interacting aggressively with metallic surfaces and exacerbating corrosion. This reaction led to the degradation of the steel and formation of MgO as corrosion products on the surface. Cross-sectional analyses (Figures 3, 4 and 6) confirmed significant Cr and Fe depletion in the vapor region, highlighting the severe impact of gaseous HCl.

In the interface region, magnesium-based inhibitors floated on the molten salt surface and preferentially reacted with corrosive compounds (e.g., H_2O). This reaction enhanced molten salt purity and effectively reduced chromium dissolution, forming protective magnesium oxides. Additionally, magnesium underwent displacement reactions with Fe^{2+} and Cr^{2+}, depositing metallic Fe and Cr layers, clearly visible in Figure 6(b_1).

In the molten salt region, continuous removal of impurities by magnesium inhibitors minimized corrosive elements (H_2O and O_2), resulting in the highest salt purity and minimal corrosion.

These corrosion processes are summarized schematically in Figure 7. All three regions shared common corrosion pathways involving chemical reactions with H_2O and O_2, producing corrosive substances such as Cl_2 and HCl, and causing dissolution of Fe and Cr into ionic forms. Variations in corrosion severity were due to differing concentrations of Cl^-, H_2O, and O^{2-}. The highest concentrations in the vapor region resulted in the most severe corrosion, while the lowest concentrations in the molten salt region resulted in minimal corrosion. Magnesium-driven displacement reactions at the interface further altered the corrosion dynamics.

5. Conclusions

This study systematically investigated the corrosion behavior of SS347H in molten NaCl-KCl-$MgCl_2$ salt under three exposure conditions: vapor, interface, and molten salt regions. The results demonstrated that corrosion severity followed the order vapor > interface > salt. The molten salt vapor region experienced the deepest corrosion due to high concentrations of volatile HCl and aggressive oxidative species, leading to substantial Cr and Fe depletion. The interface region showed notable corrosion product accumulation and deposition of Fe and Cr via magnesium-induced displacement reactions. In contrast, the molten salt region exhibited the least corrosion, benefiting from continuous impurity scavenging by magnesium, which effectively inhibited the corrosion process.

These findings highlight the importance of impurity control and corrosion inhibitors in chloride salt environments and provide valuable guidance for the safe and durable application of SS347H in next-generation concentrated solar power systems.

Author Contributions: Z.L.: writing—original draft, methodology, detection, and data analysis. H.L.: writing—review & editing, data analysis, conceptualization, and supervision. Y.W.: supervision and resources. Y.P.: supervision and resources. L.S.: supervision and resources. J.L.: writing—review and editing, and project administration. All authors have read and agreed to the published version of the manuscript.

Funding: This work was supported by the National Key Research and Development Program of China (Grant No. 2021YFB3700602).

Institutional Review Board Statement: Not applicable.

Informed Consent Statement: Not applicable.

Data Availability Statement: The raw data supporting the conclusions of this article will be made available by the authors on request.

Conflicts of Interest: The authors declare that they have no known competing financial interests or personal relationships that could have appeared to influence the work reported in this paper.

References

1. Khan, M.I. Progress in technology advancements for next generation concentrated solar power using solid particle receivers. *Sustain. Energy Technol. Assess.* **2022**, *54*, 102813. [CrossRef]
2. Alami, A.H.; Olabi, A.G.; Mdallal, A.; Rezk, A.; Radwan, A.; Rahman, S.M.A.; Shah, S.K.; Abdelkareem, M.A. Concentrating solar power (CSP) technologies: Status and analysis. *Int. J. Thermofluids* **2023**, *18*, 100340. [CrossRef]
3. Arias, I.; Cardemil, J.; Zarza, E.; Valenzuela, L.; Escobar, R. Latest developments, assessments and research trends for next generation of concentrated solar power plants using liquid heat transfer fluids. *Renew. Sustain. Energy Rev.* **2022**, *168*, 112844. [CrossRef]
4. Ding, W.; Bauer, T. Progress in Research and Development of Molten Chloride Salt Technology for Next Generation Concentrated Solar Power Plants. *Engineering* **2021**, *7*, 334–347. [CrossRef]
5. Ding, W.; Shi, H.; Xiu, Y.; Bonk, A.; Weisenburger, A.; Jianu, A.; Bauer, T. Hot corrosion behavior of commercial alloys in thermal energy storage material of molten $MgCl_2$/KCl/NaCl under inert atmosphere. *Sol. Energy Mater. Sol. Cells* **2018**, *184*, 22–30. [CrossRef]
6. Liu, Q.; Xu, H.; Yin, H.; Li, N.; Wang, W.; Li, L.; Tang, Z.; Qian, Y. Corrosion behaviour of 316 stainless steel in NaCl-KCl-$MgCl_2$ salt vapour at 700 °C. *Corros. Sci.* **2022**, *194*, 109921. [CrossRef]
7. Li, X.; Chang, L.; Liu, C.; Leng, B.; Ye, X.; Han, F.; Yang, X. Effect of thermal aging on corrosion behavior of type 316H stainless steel in molten chloride salt. *Corros. Sci.* **2021**, *191*, 109784. [CrossRef]
8. Lambrecht, M.; García-Martín, G.; De Miguel, M.T.; Lasanta, M.I.; Pérez, F.J. Corrosion study of Ni-based alloy in ternary chloride salt for thermal storage application. *Corros. Sci.* **2022**, *208*, 110673. [CrossRef]
9. Mortazavi, A.; Zhao, Y.; Esmaily, M.; Allanore, A.; Vidal, J.; Birbilis, N. High-temperature corrosion of a nickel-based alloy in a molten chloride environment—The effect of thermal and chemical purifications. *Sol. Energy Mater. Sol. Cells* **2022**, *236*, 111542. [CrossRef]
10. Fernández, A.G.; Gomez-Vidal, J.; Oró, E.; Kruizenga, A.; Solé, A.; Cabeza, L.F. Mainstreaming commercial CSP systems: A technology review. *Renew. Energy* **2019**, *140*, 152–176. [CrossRef]
11. Walczak, M.; Pineda, F.; Fernández, Á.G.; Mata-Torres, C.; Escobar, R.A. Materials corrosion for thermal energy storage systems in concentrated solar power plants. *Renew. Sustain. Energy Rev.* **2018**, *86*, 22–44. [CrossRef]
12. Bell, S.; Steinberg, T.; Will, G. Corrosion mechanisms in molten salt thermal energy storage for concentrating solar power, Renew. *Sustain. Energy Rev.* **2019**, *114*, 109328. [CrossRef]

13. Sarvghad, M.; Maher, S.D.; Collard, D.; Tassan, M.; Will, G.; Steinberg, T.A. Materials compatibility for the next generation of Concentrated Solar Power plants. *Energy Storage Mater.* **2018**, *14*, 179–198. [CrossRef]
14. Li, Y.; Wang, S.; Sun, P.; Xu, D.; Ren, M.; Guo, Y.; Lin, G. Early oxidation mechanism of austenitic stainless steel TP347H in supercritical water. *Corros. Sci.* **2017**, *128*, 241–252. [CrossRef]
15. Wang, Q.; Li, R.; Li, J. Effect of deformation temperature on microstructure and corrosion properties of hot-compressed 347H stainless steel. *Electrochem. Commun.* **2025**, *171*, 107861. [CrossRef]
16. Guo, H.; Zang, C.; Wang, Z.; Zhao, X.; Meng, Y.; Osorio, J.D.; Mehos, M. Analysis of thermal and mechanical properties with inventory level of the molten salt storage tank in central receiver concentrating solar power plants. *Appl. Therm. Eng.* **2025**, *260*, 124984. [CrossRef]
17. Mayo, C.; Batuecas, E.; Díaz, R.; Pérez, F.J. Comparative environmental assessment of two materials suited to central tower CSP technology. *Sol. Energy* **2018**, *162*, 178–186. [CrossRef]
18. Liu, Q.; Wang, Z.; Liu, W.; Yin, H.; Tang, Z.; Qian, Y. Ni-Mo-Cr alloy corrosion in molten NaCl-KCl-$MgCl_2$ salt and vapour. *Corros. Sci.* **2021**, *180*, 109183. [CrossRef]
19. Zahs, A.; Spiegel, M.; Grabke, H.J. Chloridation and oxidation of iron, chromium, nickel and their alloys in chloridizing and oxidizing atmospheres at 400–700 °C. *Corros. Sci.* **2000**, *42*, 1093–1122. [CrossRef]
20. Mohanty, B.P.; Shores, D.A. Role of chlorides in hot corrosion of a cast Fe–Cr–Ni alloy. Part I: Experimental studies. *Corros. Sci.* **2004**, *46*, 2893–2907. [CrossRef]
21. Sun, H.; Wang, J.; Li, Z.; Zhang, P.; Su, X. Corrosion behavior of 316SS and Ni-based alloys in a ternary NaCl-KCl-$MgCl_2$ molten salt. *Sol. Energy* **2018**, *171*, 320–329. [CrossRef]
22. Guo, L.; Liu, Q.; Yin, H.; Pan, T.J.; Tang, Z. Excellent corrosion resistance of 316 stainless steel in purified NaCl-$MgCl_2$ eutectic salt at high temperature. *Corros. Sci.* **2020**, *166*, 108473. [CrossRef]
23. Sun, H.; Zhang, P.; Wang, J. Effects of alloying elements on the corrosion behavior of Ni-based alloys in molten NaCl-KCl-$MgCl_2$ salt at different temperatures. *Corros. Sci.* **2018**, *143*, 187–199. [CrossRef]

Disclaimer/Publisher's Note: The statements, opinions and data contained in all publications are solely those of the individual author(s) and contributor(s) and not of MDPI and/or the editor(s). MDPI and/or the editor(s) disclaim responsibility for any injury to people or property resulting from any ideas, methods, instructions or products referred to in the content.

materials

MDPI

Article

Composition-Dependent Creep Resistance and Strain Rate Sensitivity of BCC Mg-Sc Alloy Studied via Nano-Indentation on Diffusion Couple

Chenyue Liu, Guanglong Xu * and Fuwen Chen

College of Materials Science and Engineering, Nanjing Tech University, Nanjing 211816, China
* Correspondence: guanglongxu@njtech.edu.cn

Abstract

Mg-Sc body-centered cubic (BCC) phase-structured alloys not only exhibit superior room-temperature ductility and quasi-isotropic deformation behaviors compared to conventional hexagonal close-packed (HCP) Mg alloys in mechanical applications, but they also demonstrate a shape-memory effect that is applicable to intelligent devices. Due to the introduction of a dual-phase microstructure feature, the unveiled strengthening/toughening mechanism, and the potential benefit of Sc alloying in BCC creep deformation, it is necessary to investigate the composition and time-dependent creep behaviors of BCC Mg-Sc alloys, such as creep resistance and strain rate sensitivity at room temperature, through nano-indentation on the Mg-Sc diffusion couple. A critical finding is that as the Sc content increases from 23.01 at.% to 33.56 at.%, the BCC Mg-Sc alloy exhibits a progressive enhancement in creep resistance at room temperature, evidenced by the creep stress exponent (n) rising from 49.02 to 66.22. Furthermore, the strain rate sensitivity (m) increases from 0.02 at 26.94 at.% Sc to 0.11 at 32.63 at.% Sc, along with the Sc composition gradient. These phenomena can be attributed to the formation of ordered structures with the increasing Sc concentration, which introduce short-range local barriers to dislocation motion, as confirmed through atomic-scale microstructural analysis.

Keywords: Mg-Sc BCC alloy; diffusion couple; nano-indentation; creep resistance; strain rate sensitivity

1. Introduction

Magnesium alloys have received significant interest in the field of engineering due to their good castability [1], high specific stiffness [2,3], and superior electrical conductivity [4,5], with wide applications in the 3C and transportation sectors [6–8]. Rare-earth elements, such as Gd [9], Y [10], Nd [11], etc., are commonly employed for alloying the hexagonal close-packed (HCP) lattice of Mg and introducing thermally stable precipitates [10,12] to enhance heat resistance and creep resistance [9–11]. Nevertheless, HCP-structured Mg has an intrinsic problem of inadequate plasticity deformation [13].

Recent years have witnessed a growing interest in body-centered cubic (BCC)-structured Mg alloys. Mg-Li BCC alloys have good performance in terms of ductility but suffer from inherently low creep resistance [14]. Apart from Li, only Sc can be solid-solutioned with Mg atoms and shaped into a BCC solid solution with a broad composition across a varying range [15]. Mg-Sc alloys have been found to be functional materials featuring a shape-memory effect, strain glass, and biocompatibility [16–19]. Meanwhile, as structural materials, they open up a novel pathway of dual-phase strengthening, similar to Ti alloys, beyond dispersed precipitation hardening [20–24]. Mg-Sc BCC matrix alloys are as light as organic-based materials with a density of ~2 g· cm^{-3}, possess a nearly temperature-independent and low Young's modulus of ~20–23 GPa from room temperature down to ~100 K [18], a Vickers hardness of 231.5 Hv [22], a yield strength of ~200–270 MPa [19], an ultimate tensile strength of 310 MPa, an elongation of 28.8% [25], and a long fatigue life, and can overcome the tradeoff between the strength and ductility of conventional Mg alloys [19].

In addition, Sc lies in the IIIB group of the periodic table, the same as rare-earth elements. The effect of Sc on creep resistance has been studied for HCP-structured Mg alloys [26–30]. The improvement in creep resistance by Sc alloying is associated with the pinning of Sc-containing short-range particles, which influences the cross-slip of dislocations [31–33], and a low tendency for Sc diffusion, which imposes drag on migrating grain boundaries [34]. However, the effect of Sc alloying on the creep resistance of BCC-structured Mg alloys remains limited. In particular, creep displacement, the creep stress exponent, and strain rate sensitivity are critical parameters for evaluating creep deformation behaviors [35]. As such, a systematic investigation into the composition dependence of creep resistance-relevant parameters for Mg-Sc BCC alloys is necessary.

The employment of a conventional uniaxial tensile creep test to study creep deformation behaviors at the macro-scale is time-consuming and cost-intensive [36]. This has advanced the adoption of nano-indentation as a novel approach to investigate creep deformation behaviors at the micro-scale, enabling the evaluation of both time-dependent creep deformation and strain rate sensitivity [37], and this technique has been successfully implemented to study these phenomena in pure Mg and Mg-X (X = Al, Ca) systems [38–40]. Recent advances include the combination of nano-indentation with a solid-state diffusion couple [41,42] (which has the ability to introduce a continuous composition gradient) to serve as a high-throughput method to screen for composition-dependent mechanical properties [43].

Building upon these technical developments, the present work employs a synergy of the solid-state diffusion couple and nano-indentation to address the lack of data on composition-dependent creep behavior in BCC Mg-Sc systems at room temperature. The variations in the creep displacement, creep stress exponent (n), and strain rate sensitivity (m) with Sc contents are evaluated. It is noteworthy that this work focuses on the intrinsic creep resistance and the intrinsic strain rate sensitivity purely as influenced by the chemical composition of Sc at room temperature. Other effects of thermal-activated deformation [44], such as grain boundary sliding and grain boundary diffusion at elevated temperatures, are not involved.

2. Materials and Methods

2.1. Fabrication of Diffusion Couple

A Mg-Sc diffusion couple (see the schematic drawing in Figure 1) was synthesized through solid-state diffusion bonding. The end members were pure Mg (99.95 wt.%) and Sc (99.9 wt.%) blocks (Jinyu Sunshine Metallic Materials Co., Ltd., Beijing, China), which were homogenized at 450 °C (723 K) and 800 °C (1073 K), respectively, for 48 h under an argon atmosphere. Rectangular specimens with the size of 10 mm × 10 mm × 5 mm were sectioned from homogenized ingots using wire electrical discharge machining, followed by mechanical polishing to achieve mirror-like surfaces. The polished Mg and Sc blocks were mechanically clamped in a stainless-steel fixture to ensure intimate interfacial contact. The assembled diffusion couple was sealed within an evacuated quartz tube, backfilled with high-purity Ar. It was then subjected to annealing at 630 °C (903 K) for 72 h, also in an Ar atmosphere, to form an intermediate BCC phase, the diffusional interfaces of Mg-rich HCP/BCC and BCC/Sc-rich HCP, as well as the chemical composition gradients in individual phases. The annealing temperature and time were set based on the previous work of the present authors [25,45,46], ensuring a thickness of the generated BCC layer of more than 1mm, a sigmoid shape of composition–distance profiles, and an appropriate BCC grain size distribution. Annealing in Ar prevented oxidation and other contamination, minimized the volatilization of Mg via Le Chatelier's principle, and ensured microstructural stability. Post-diffusion processing involved longitudinal sectioning of the annealed couple via precision wire electrical discharge machining, followed by sequential polishing using colloidal silica suspension. The microstructural features of phases and grains in the diffusion couple were characterized using a TESCAN S8000 GMH ultra-high-resolution scanning electron microscope (SEM, TESCAN, BrNo, Czech) equipped with an Oxford Instruments Nordlys Electron Backscatter Diffraction (EBSD, Oxford Instruments, Oxford, UK) detector. EBSD mapping and backscattered imaging were performed at an acceleration voltage of 20 kV and a current of 20 nA. Based on the inherent characteristics of the alloy, the scanning step size was set to 1.4–2.5 μm. The EBSD data were analyzed with Channel 5 software (Version 5.12.74.0). The crystal structure of pure Mg served as the reference pattern for the HCP phase, while the MgSc crystal structure was selected as the calibration standard for the BCC phase. Local chemical compositions were quantitatively point-to-point mapped via energy-dispersive X-ray spectroscopy (EDS, Oxford Instruments, Oxford, UK) with the same acceleration voltage on SEM. The mapping direction was along the diffusion direction.

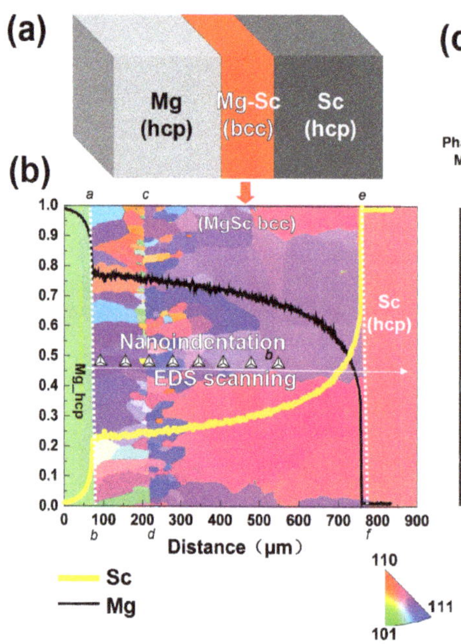

Figure 1. (**a**) Schematic drawing of sandwich three-phase structure in the diffusion couple. The BCC phase in red was generated in annealing. (**b**) EBSD orientation map of diffusion couple superposed with EDS component–distance curves (yellow curve is for Sc composition and black curve is for Mg composition), and schematic drawing of nano-indentation positions showing using triangles (not experimental observation). (**c**) Back-scattered SEM image and phase identification map. The phase boundaries and initial wielding interface are also highlighted.

2.2. Nano-Indentation Creep Deformation and Continuous Stiffness Tests

The creep resistance evaluation was performed using Nano Indenter G200 (KLA Instruments, Phoenix, AZ, USA). We tested 20 indentation points stepwise at 20 different positions with different Sc compositions. These 20 indentation points were arranged in 2 rows parallel to the diffusion direction (composition gradient of Sc) to minimize interference from surface defects and grain boundaries. A spacing of 50 μm was maintained between adjacent indentation positions to avoid elastoplastic interactions. Based on the reference nano-indentation parameters for other Mg diffusion couples [12,47] and pre-test parameter tuning, a load-controlled mode was employed with a peak load of 10 mN (which can achieve a sufficient plastic zone depth to probe bulk-like creep behavior) applied at a constant loading rate of 1 mN/s. The maximum load was held for 100 s to ensure steady-state creep establishment. Subsequently, the specimen was unloaded with a 75 s thermal drift correction hold segment at 90% of the peak load during the unloading.

For strain rate sensitivity characterization, four microregions with different Sc compositions were probed via the continuous stiffness measurement (CSM) methodology to acquire hardness–depth profiles under varying strain rates (0.01 s^{-1}, 0.05 s^{-1}, and 0.1 s^{-1}). A constant penetration depth of 1000 nm was maintained for all tests. The stabilized hardness values were determined by averaging data from the region (400–1000 nm depth) to minimize surface effects.

After a nano-indentation test, the chemical composition of Sc near the indentation was detected using low-voltage EDS point mapping. And transmission electron microscopy (TEM) observation was performed including bright-field (BF) imaging and selected area electron diffraction (SAED). The focused ion beam (FIB) was applied to prepare TEM specimens from two different near-indentation regions, but with significant differences in chemical composition. The ordering/disordering transition in BCC phases was addressed via TEM using a Talos F200X microscope (Thermo Fisher Scientific, BrNo, Czech) at an accelerating voltage of 200kV.

2.3. Data Analyses of Nano-Indentation Creep Deformation Test

The creep response of nano-indentation can be quantitatively described by the stress and time during the load holding stage, that is, following the Norton power-law model [48].

$$\sigma = \frac{P}{S} \tag{1}$$

Among them, P represents the instantaneous load, and S represents the instantaneous contact area between the indenter and the sample during the indentation process.

$$\dot{\varepsilon} = A\sigma^n \exp\left(-\frac{Q}{RT}\right) \tag{2}$$

In Equation (2), $\dot{\varepsilon}$ is the strain rate during holding, σ is the instantaneous stress, A is a constant, n is the stress exponent, Q is the creep activation energy, R is the universal gas constant (8.314 $J{\cdot}mol^{-1}{\cdot}K^{-1}$), and T is the temperature in Kelvin. Creep activation energy, Q, reflects the minimum energy required for atoms to cross the potential barrier during creep. Its value is usually related to the diffusion mechanism and characterizes the energy barrier required for atoms to migrate from high-energy positions to low-energy positions. In this study, all nano-indentation creep tests were conducted at room temperature (298 K). Thus, T = 298 K was used in the analysis.

Meanwhile, the strain rate $\dot{\varepsilon}$ can also be expressed by indentation depth, as in Equation (3).

$$\dot{\varepsilon} = \frac{1}{h}\frac{dh}{dt} \tag{3}$$

In this nano-indentation experiment, the Bercovich triangular pyramid indenter was used. Therefore, under ideal conditions, the relationship between its contact area and indentation depth is shown in Equation (4).

$$S = 24.56h^2 \tag{4}$$

When substituting Equations (1), (3) and (4) into (2), integrating the stress σ and time t, and then taking the logarithm, the relationship between creep stress and creep time during the indentation holding process can be obtained as shown in Equation (5).

$$-n\ln(\sigma) = \ln(t) + \ln B - Q/RT \tag{5}$$

where B is constant and $ln(\sigma)$ and $ln(t)$ show a linear relationship. The slope of the linear regression plot is associated with the negative reciprocal of the creep stress exponent n.

2.4. Data Analyses of Nano-Indentation Continuous Stiffness Measurement

The strain rate sensitivity response of nano-indentation is derived. The material hardness H measured by the nano-indentation can be expressed as

$$H = C\dot{\varepsilon}^m \tag{6}$$

where C is a constant, $\dot{\varepsilon}$ is the indentation strain rate ($\dot{h}/h$), which can be expressed as the loading rate divided by the load value, and m is the strain rate sensitive factor, whose value is always greater than and close to 0. Take the logarithm of both sides of Equation (6),

$$\ln(H) = m \cdot \ln(\dot{\varepsilon}) + \ln(C) \tag{7}$$

In this way, the linear relationship between the logarithm of hardness and the logarithm of strain rate can be obtained, and its slope is the strain rate sensitive factor m.

$$m = \frac{d\ln(H)}{d\ln(\dot{\varepsilon})} \tag{8}$$

3. Results and Discussion

3.1. Microstructural Characterization

Figure 1a schematically shows, and the phase map in Figure 1c experimental evidences, the HCP/BCC/HCP sandwich structure of the diffusion couple after annealing at 630 °C (903 K) for 72 hours, which contains three phases. On the left is the end-member of the Mg-rich HCP phase, and on the right is the end-member of the Sc-rich HCP phase. Both HCP phases exist before annealing. In between, a layer of the BCC phase has been generated.

Figure 1b shows the EBSD orientation mapping projected by the colors of the inverse pole figure (IPF). The Mg-rich HCP phase (on the left, in green) and the Sc-rich HCP phase (on the right, in pink) show extremely coarsened grains. There are no grain boundaries in the figure horizon. The BCC phase exhibits an increasing grain size from the original weld interface towards the left and right sides, as the initial weld interface can serve as a preferential heterogeneous nucleation site. However, the initial weld interface was generally smeared out during annealing diffusion. Thus, the vicinity of the original weld interface produces fine equiaxed grains, and a progressive coarsening trend emerges with increasing distance from the interface, accompanied by preferential grain elongation parallel to the diffusion direction. Due to the effect of heterogeneous nucleation on the initial welding interface, which is dominant in grain size distribution, it is not proof-positive that the grain size in the BCC region is related to the varying Sc contents.

Figure 1b also exhibits the EDS-detected chemical composition–distance profiles. The profile of Sc is in yellow, and Mg is in black. Thanks to the reaction diffusion between Mg and Sc, not only is the BCC phase generated but also the continuous chemical composition gradients are formed in individual phases, especially in the BCC layer. There is a discontinuous composition transition taking place at the Mg-rich HCP/BCC interface, where the Sc composition abruptly increases from 11.66 at.% to 22.78 at.%, indicating the solid solubility of 630 °C (903 K) on the phase diagram. Figure 1b also schematically shows the direction of EDS scanning, which is parallel to the composition gradient of Sc. Thus, the nano-indentations are implemented along the same direction, such that the composition dependence of creep deformation behaviors can be figured out.

Figure 1c shows a backscattered SEM image of the diffusion couple, together with the phase identification image in the inset and the marked interfaces. The SEM image evidences a well-developed surface without cracks and the well-developed phase boundaries between HCPs and BCC. Although the grain size morphology changes with diffusion distance, there is neither cracking nor are there discontinuous defects inside the BCC phase. We can identify the trace of the initial weld interface only by SEM observation, indicating a high quality of diffusion bonding to prepare the diffusion couple. The generated phase interfaces of Mg-rich HCP/BCC and BCC/Sc-rich HCP are almost perpendicular to the composition gradient, indicating that a planar diffusion has taken place in the diffusion couple, such that the chemical compositions perpendicular to the Mg/Sc inter-diffusion are almost the same [49,50], and the nano-indentation perpendicular to the phase interface is along the chemical composition gradient of Sc.

Considering the experimental results showing that (1) the BCC layer with a single-phase thickness ≈ 650 μm is enough to implement dozens of nano-indentations with the necessary spacing, (2) there is a big enough grain size to avoid the effect of a grain boundary, (3) there is a composition range of Sc from ~23% to ~40% with an appropriate (not too steep) spatial gradient, (4) there is a high metallurgical quality of the BCC phase without cracking and discontinuous defects, the prepared BCC layer is qualified to implement a nano-indentation test inside the individual grains and then further to explore the composition dependence of creep behaviors of Mg-Sc BCC alloys at room temperature.

3.2. Nano-Indentation Creep Behavior

Figure 2 displays the load–displacement responses obtained by nano-indentation testing on BCC-structured microregions with varying Sc compositions from 23.01 to 33.56 at.% under 10 mN maximum load. The selection of this target composition range and the corresponding area on the diffusion couple was considered to ensure the appropriate composition gradient to track the continuous variation in creep behaviors with Sc contents, and the spacing of 50 μm between two individual nano-indentations was set to avoid elastoplastic interreaction. However, the region with 35 at.%~60 at.% Sc of the BCC phase was dropped in this work. The chemical composition gradient of Sc was too steep to maintain the high spatial resolution of the chemical composition and the uninterrupted creep properties (see Figure 1b). Furthermore, in the Sc content lower than 23 at.%, the HCP phase can be detected in the diffusion couple. There are plateaus during the 100 s dwell period in all the tested samples (see the holding-stage characteristics in individual insets) demonstrating the occurrence of creep deformation. The creep displacement Δh (defined as depth variation during load holding) fluctuates and decreases from 17.26 nm to

11.75 nm as the Sc content increases from 23.01 at.% to 33.56 at.%. It is noteworthy that the nano-indentation penetration depth is measured at ~500nm, as shown in Figure 2. The size of the elastoplastic affected zone introduced by the indenter is assumed to be 5–10 times the penetration depth [51–53], with the maximum size less than 5 μm. In comparison with the BCC grain size in Figure 1, the dislocation movement in the elastoplastic affected zone will hardly be blocked by the grain boundary. The variation in creep displacement is primarily attributed to Sc alloying.

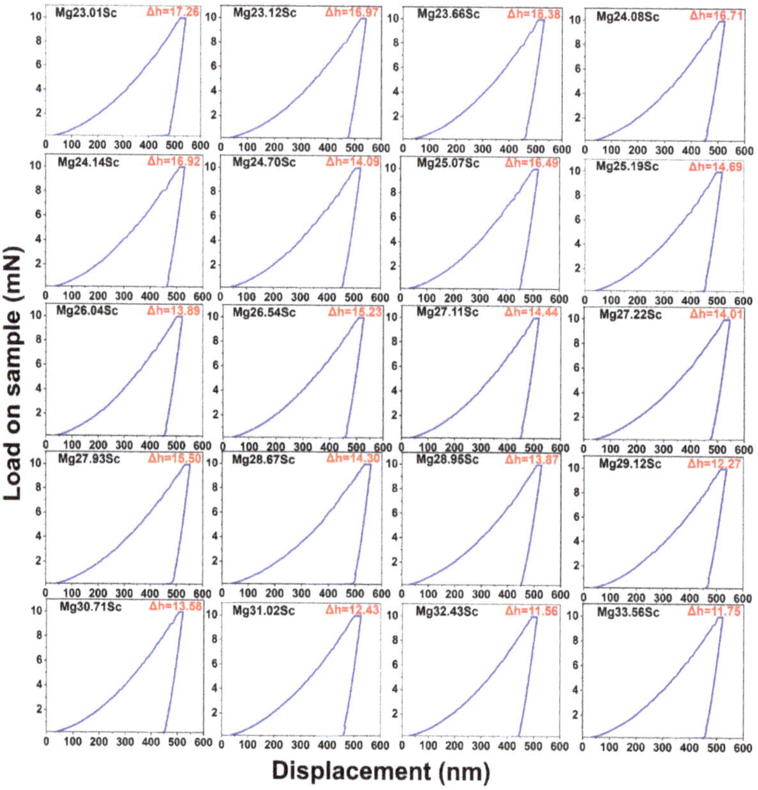

Figure 2. Load–displacement curves obtained through nano-indentation testing on BCC-structured microregions with varying Sc compositions ranging from 23.01 to 33.56 at.%.

There is an empirical formula in conventional tensile/compression experiments to describe creep displacement variation with time.

$$u_c = u_0 + a(t - t_0)^b + kt \tag{9}$$

where u_c is the creep displacement, u_0 is the displacement at the beginning of the test, t is time, t_0 is the start time of creep, and a, b, and k are the fitting parameters. In a nano-indentation creep experiment, the loading–unloading curve is not 'smooth' (not differentiable everywhere). Since we only focus on the holding stage for creep, the start time and displacement of creep should be extracted from individual nano-indentation curves, and then be subjected to numerical fitting and renormalized to zero. By renormalization, Equation (9) is simplified into

$$h_c = at^b + kt \tag{10}$$

where h_c is the nano-indentation creep depth and u_0 and t_0 will be normalized to zero.

Based on Equation (10), a power function regression analysis was conducted on the creep displacement–time relationship. In Figure 3, the experimental displacement–time curves are compared with the fitting curves using Equation (10). The experimental curves were obtained under different composition conditions after zero-point correction of the displacement–time data, holding the load for 100 s. All the experimental curves show the features of creep deformation, i.e., the deformation can be divided into two stages: an initial transient creep and subsequent steady-state creep. In the initial stage, the displacement shows a rapid increase with time. Subsequently, the deformation rate gradually slows down, and the creep curves decay. After entering the stable stage, the displacement turns to a linear increase. These behaviors are well described by Equation (10). The fitted curves and the experimental curves show good consistency with the fitted correlation coefficient R^2 values higher than 0.98. The specific parameters (the fitting constants of a, b, and k and the corresponding R^2 values) are detailed in Table 1. Fitted parameters a, b, and k and correlation coefficients R^2 were obtained by fitting creep displacement–time curves for different Sc contents using Equation (10).

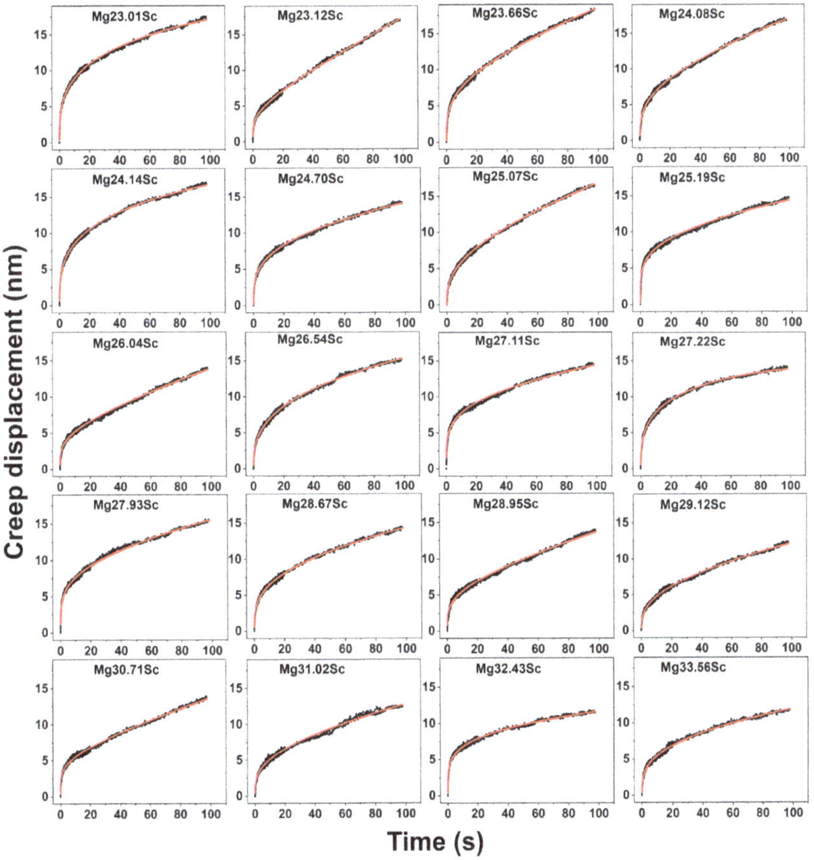

Figure 3. Creep displacement–time curves and fitted curves for BCC-structured MgSc alloys with different compositions.

Table 1. Fitted parameters *a*, *b*, *k* and correlation coefficients R^2 obtained by fitting creep displacement–time curves for different Sc contents using Equation (10).

No.	Sc at.%	a	b	k	R^2
1	23.01	4.3149	0.3175	−0.0147	0.9946
2	23.12	3.4316	0.2821	0.0158	0.9935
3	23.66	1.9768	0.3581	0.0189	0.9932
4	24.08	2.6334	0.2431	0.0586	0.9923
5	24.14	3.0238	0.3142	0.0428	0.9937
6	24.70	4.4832	0.2188	0.0221	0.9810
7	25.07	2.4008	0.3014	0.0320	0.9903
8	25.19	2.7927	0.2506	0.0495	0.9884
9	26.04	3.0589	0.2039	0.0589	0.9902
10	26.54	4.1954	0.2437	0.0271	0.9926
11	27.11	4.5016	0.2236	0.0183	0.9843
12	27.22	4.1373	0.3185	−0.0115	0.9959
13	27.93	2.9581	0.3583	−0.0067	0.9953
14	28.67	3.7173	0.2733	0.0547	0.9945
15	28.95	3.6620	0.3489	−0.0441	0.9937
16	29.12	2.4741	0.3439	0.0481	0.9969
17	30.71	4.0704	0.2161	0.0065	0.9854
18	31.02	2.6011	0.2266	0.1001	0.9959
19	32.43	3.0927	0.3041	0.0183	0.9941
20	33.56	2.5880	0.3196	0.0068	0.9901

Figure 4 depicts the creep stress exponent (n) with the composition of Sc, which was derived using Equation (5). The calculated n values increase from 49.02 to 66.22 as Sc% rises from 23.01 at.% to 33.56 at.%. The enhancement of n correlates with improved creep resistance, since it indicates a greater inherent resistance to time-dependent plastic flow under constant load conditions.

Figure 5 exhibits a good linear Sc composition dependence of creep displacement and the stress exponent. Considering the random error sensitivity of nano-indentation [54], the small error fluctuations among individual tests and the good performances of linear regressions indicate the repeatability and reliability of the experimental results. Figure 5a quantifies the variation in creep displacement (Δh) under a 10 mN holding load with Sc contents. It also describes the composition-dependent creep resistance of BCC-structured MgSc alloys. The experimental Δh exhibits an inverse and linearly proportional relation to the Sc content, yielding a slope of −0.54 nm/at.% and $R^2 = 0.77$. This trend quantitatively confirms progressively inhibited creep activity with Sc enrichment in the BCC phase.

Figure 5b reveals a proportional increase in stress exponent (n) with Sc content, following a linear dependence of the slope: 2.44 and $R^2 = 0.72$. The monotonic increase in n from 49.02 to 66.22 also demonstrates that the addition of Sc to the BCC matrix can effectively strengthen time-dependent deformation resistance. This correlation aligns with prior investigations [34] showing that Sc additions intensify elastic interactions between solute atoms and dislocations, thus enhancing the comprehensive mechanical properties of the alloy.

There is a specific connection between the stress exponent n and the creep deformation mechanism [55]. All n values varying with Sc contents are greater than 3 in this work, indicating that dislocation movement plays a significant role in creep behavior [56], and the grain boundary sliding and grain boundary diffusion creep mechanisms are not applicable under the current experimental conditions. As is mentioned in the previous paragraph, there are differences in magnitude between the size of indentation (elastoplastic affected zone) and the grain size in this work. The probability of the indentation arising mainly in the grain boundary area was low enough. Most of the indentations were implemented in the intragranular area. Therefore, the effect of grain boundaries on the creep deformation behaviors is less important in this work.

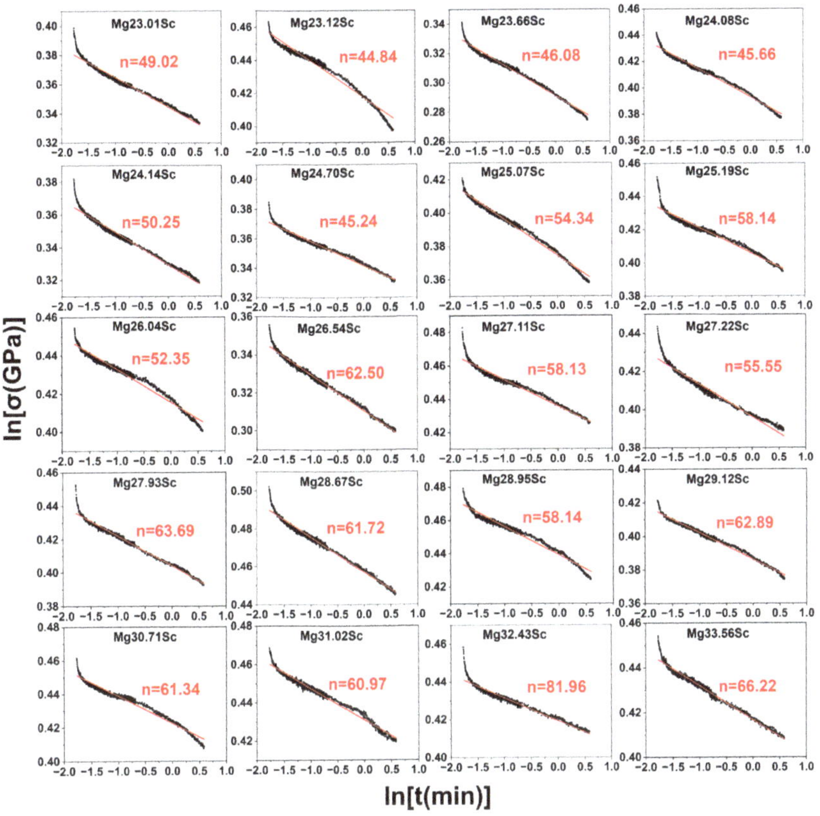

Figure 4. $\ln(\sigma)$-$\ln(t)$ relationship curves and creep stress exponent (n) for BCC-structured MgSc alloys with different compositions.

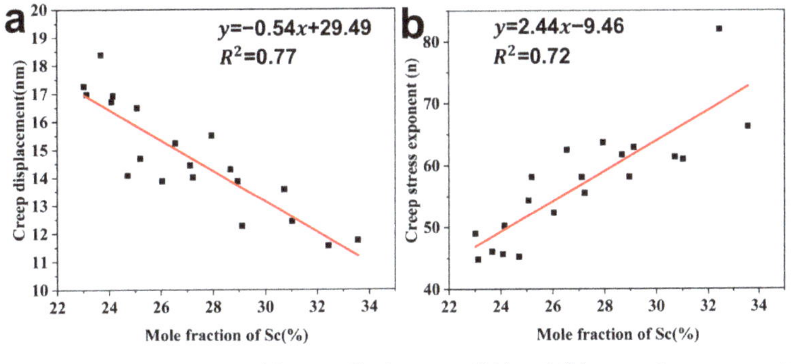

Figure 5. The variations in (**a**) creep displacement (Δh) and (**b**) creep stress exponent (n) with Sc content.

Figure 6a shows the BF image of the BCC-structured MgSc alloy at an Sc content of 21.67 at.%. The SAED pattern corresponding to Figure 6a is shown in Figure 6b, which proves that this region is a single disordered BCC phase. Figure 6c,d show the BF images and SAED at Mg 32.66 at.% Sc. It can be seen that the additional spots appear under the [100] zone axis and are located at 1/2{200}, indicated by yellow arrows. These additional spots suggest that the formation of the BCC-B2 ordered structure occurs at Mg 32.66 at.% Sc. This is also consistent with what was mentioned in previous studies: the BCC-B2 phase exists when the Sc content is within the range of 35–50 at.% Sc composition [16,25].

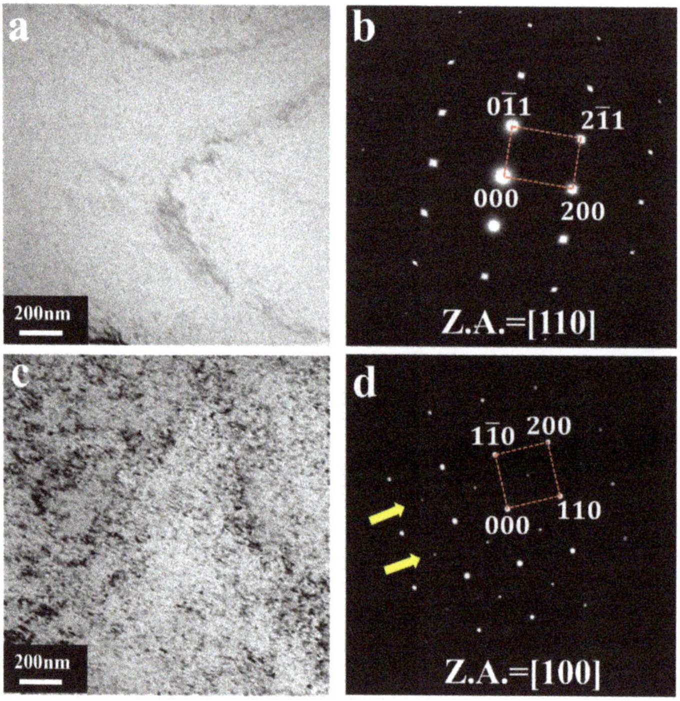

Figure 6. TEM test results of microregions. (**a**) BF image. (**b**) SAED at Mg 21.67 at.% Sc. (**c**) BF image. (**d**) SAED at Mg 32.66 at.% Sc. The yellow arrows highlight the additional reflection array at $\frac{1}{2}\{200\}$, indicating the formation of B2 ordered structure.

With the SAED patterns, we estimated the lattice parameter a = 3.581 Å at Sc = 21.67 at.% and a = 3.690 Å at Sc = 32.66 at.%. It is evidenced that the alloying of Sc results in BCC lattice expansion. The random solid solution of Sc leads to local lattice distortion, which hinders the movement of dislocations and enhances creep resistance. Moreover, the change in lattice parameter results in a varied magnitude of Burgers vector, then influences the pre-exponential factor A and flow stress in the equation for the creep rate (Equation (2)). On the other hand, with an increasing Sc content, there is atomic-scale ordering of B2 taking place in the disordered BCC matrix, as confirmed by SAED observations shown in Figure 6d. These ordered precipitates possess a significantly higher energy barrier for dislocation penetration compared to the disordered matrix, as predicted by theories of precipitate strengthening [57]. Consequently, dislocations interacting with these precipitates experience increased drag forces, necessitating thermally activated climb processes to bypass obstacles [58,59]. This climb-controlled bypass mechanism inherently reduces dislocation mobility, leading to the observed decrease in creep displacement and the significant enhancement in overall creep resistance of the alloys.

3.3. Nanometer Indentation Strain Rate Sensitivity

Figure 7 displays the hardness–depth profiles of the samples measured via the CSM method under various strain rates (0.01 s^{-1}, 0.05 s^{-1}, and 0.1 s^{-1}). At shallow indentation depths (<400 nm), the hardness values exhibited significant fluctuations with increasing depth. However, the curves were gradually stabilized when the indentation depth exceeded 400 nm. Consequently, the average hardness values within the 400–1000 nm depth range were selected as the experimentally determined hardness. The results reveal an increasing trend in hardness with higher strain rates, indicating notable strain rate sensitivity of hardness in BCC-structured MgSc alloys. Furthermore, the influence of the strain rate on hardness becomes more pronounced with an increasing Sc content, as evidenced by the growing discrepancy between the curves. In terms of composition dependence, the stabilized hardness values remained approximately 2 GPa across all tested compositions, suggesting that Sc additions within this compositional range did not effectively enhance the hardness of Mg-Sc alloys.

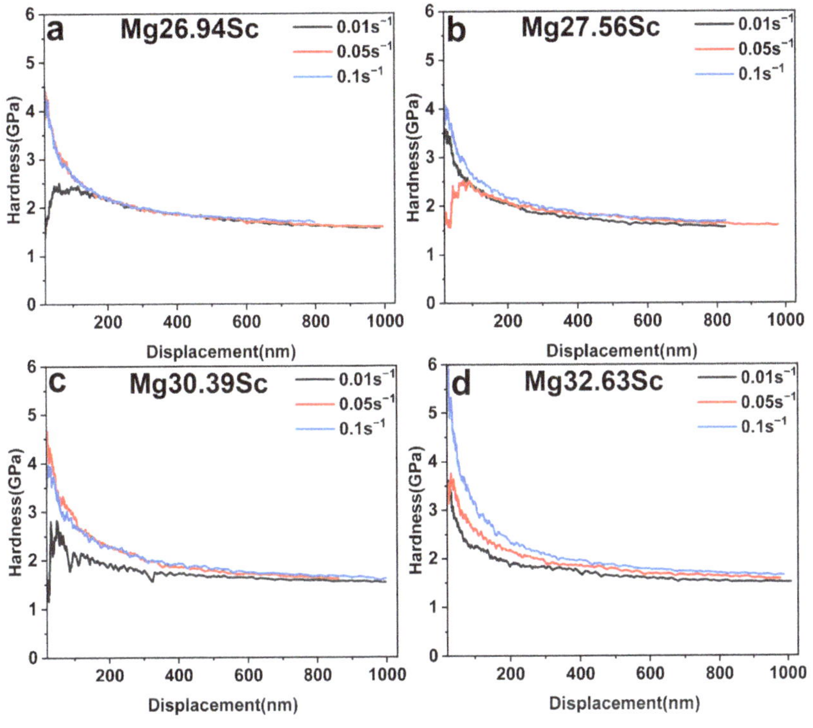

Figure 7. The hardness–depth curves of BCC-structured MgSc alloys with different compositions of Sc measured at various strain rates: (**a**) Mg 26.94 at.% Sc, (**b**) Mg 27.56 at.% Sc, (**c**) Mg 30.39 at.% Sc, (**d**) Mg 32.63 at.% Sc.

Figure 8 shows the strain rate sensitivity index (m), as evaluated by the ln (H)-ln ($\dot{\varepsilon}$) relationship with Equation (8). With the increase in Sc content from 26.94 at.% to 32.63 at.%, the m value increases from 0.02 to 0.11. The creep stress exponent (n) ascertained by nano-indentation creep and the strain rate sensitivity index (m) ascertained by CSM were obtained individually. The values of m and n from individual tests approximately satisfy the theoretical constraint of $m = 1/n$, also indicating the reliability of the results.

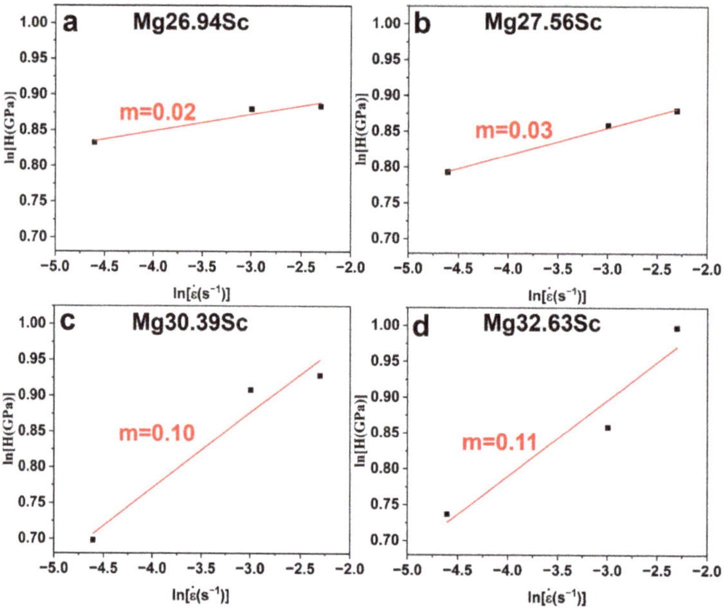

Figure 8. ln(H)-ln(ε) relationship curves and strain rate sensitivity index (m) of BCC-structured MgSc alloys with different compositions: (**a**) Mg 26.94 at.% Sc, (**b**) Mg 27.56 at.% Sc, (**c**) Mg 30.39 at.% Sc, (**d**) Mg 32.63 at.% Sc.

The present authors have studied the correlation between solvent diffusion and creep properties [12] and measured the diffusion coefficients in BCC-structured Mg-Sc alloys [45,46]. Sc acts as a diffusion-slowing solute in BCC Mg, leading to a decrease in the solvent diffusion of Mg atoms in Mg-Sc BCC solid solution, which may impose additional drag or friction on the movement of dislocations when the content of Sc is increased. More importantly, the enhanced m with the Sc content is not linear, showing a sudden increase at around 30 at.% Sc. According to the Mg-Sc binary-phase diagram, this may correlate with the formation of the chemically ordered B2 precipitates demonstrated in Figure 6. These ordered B2 precipitates act as localized barriers and obstacles to dislocation motion [60]. The effect amplifies as the strain rate increases.

4. Limitations

While this study provides valuable insights into the Sc composition dependence of creep behavior in bcc Mg-Sc alloys, several limitations warrant acknowledgment:

1. Only the material intrinsic creep resistance, the stress exponent, and the intrinsic strain rate sensitivity at room temperature, which are purely influenced by the chemical composition of Sc, have been studied. The creep deformation behaviors at elevated temperatures, which are activated by thermal physics, and the mechanisms of grain boundary sliding and grain boundary diffusion have not been considered yet. They are left for future works.
2. Strain rate range constraint: The nano-indentation creep tests were conducted within a defined strain rate range of 0.01 s^{-1} to 0.1 s^{-1}. While appropriate for probing local creep resistance, this relatively narrow window may restrict the comprehensive quantification of the strain rate sensitivity exponent (m), potentially limiting the resolution of its composition-dependent evolution.

3. Microstructural correlation gap: The conclusions regarding strengthening mechanisms are based solely on mechanical property measurements (n, m). The absence of complementary microstructural characterization (e.g., TEM for dislocation substructures or precipitate analysis) prevents a definitive mechanistic interpretation of the observed composition–property relationships. Future work integrating microstructural probes is essential to elucidate the underlying mechanisms driving the enhanced creep resistance.

5. Conclusions

The composition-dependent creep resistance and strain rate sensitivity of BCC-structured Mg-Sc alloys were investigated through a nano-indentation creep deformation test and continuous stiffness measurement on the Mg-Sc binary diffusion couple. The key findings are summarized as follows:

1. Sc significantly improves the room-temperature creep resistance of BCC-structured MgSc alloys. The creep displacement Δh decreases from 17.26 nm to 11.75 nm, while the stress exponent n increases from 49.02 to 66.22 as the Sc content rises from 23.01 at.% to 33.56 at.%. The composition variations in creep displacement and creep stress exponent indicate a better creep resistance of a Mg-Sc binary alloy at high-Sc compositions in comparison with low-Sc compositions. TEM characterization reveals that this enhancement originates from Sc-induced ordered structures, which act as practical barriers to dislocation motion through dynamic drag effects during creep deformation.
2. While the stabilized hardness remains at approximately 2 GPa across all compositions, the strain rate sensitivity index m increases from 0.02 to 0.11 with a higher Sc content at room temperature. This phenomenon is attributed to the proliferation of short-range obstacles (e.g., ordered structure) that amplify the activation barriers for dislocation glide under strain rate variations.

Author Contributions: Conceptualization, G.X. and F.C.; methodology, G.X.; validation, G.X. and F.C.; investigation, C.L.; resources, G.X.; data curation, G.X.; writing—original draft preparation, C.L.; writing—review and editing, G.X. and F.C.; visualization, C.L.; supervision, G.X. and F.C.; project administration, F.C.; funding acquisition, G.X. All authors have read and agreed to the published version of the manuscript.

Funding: The work was sponsored by the National Natural Science Foundation of China [Grant Nos. 51701094, 50731002, 51911540474] and the Natural Science Foundation of Jiangsu Province, China [Grant No. BK20171014] as a part of Mg-based alloy design. G.X. acknowledges the financial support from Jiangsu Provincial Innovation Support Program's "Belt and Road" Innovation Cooperation Project [Grant No. BZ2023006].

Institutional Review Board Statement: Not applicable.

Informed Consent Statement: Not applicable.

Data Availability Statement: The data presented in this study are openly available in Mendeley Data at 10.17632/yr69v4x52k.1.

Conflicts of Interest: The authors declare no conflicts of interest.

References

1. Lv, H.; Tan, J.; Geng, T.; Wang, H.; Shi, H.; Lin, Y.; Jiang, B.; Eckert, J. Advances in high-strength and high-thermal conductivity cast magnesium alloys: Strategies for property optimization. *J. Alloys Compd.* **2025**, *1029*, 180843. [CrossRef]
2. Friedrich, H.; Schumann, S. Research for a "new age of magnesium" in the automotive industry. *J. Mater. Process. Technol.* **2001**, *117*, 276–281. [CrossRef]

3. Zhang, J.; Chen, C.; Cheng, W.; Bian, L.; Wang, H.; Xu, C. High-strength $Mg_{93.96}Zn_2Y_4Sr_{0.04}$ alloy with long-period stacking ordered structure. *Mater. Sci. Eng. A* **2013**, *559*, 416–420. [CrossRef]
4. Bamberger, M.; Dehm, G. Trends in the development of new Mg alloys. *Annu. Rev. Mater. Res.* **2008**, *38*, 505–533. [CrossRef]
5. Yang, J.; Zhu, Z.; Han, S.; Gu, Y.; Zhu, Z.; Zhang, H. Evolution, limitations, advantages, and future challenges of magnesium alloys as materials for aerospace applications. *J. Alloys Compd.* **2024**, 176707. [CrossRef]
6. Froes, F.; Eliezer, D.; Aghion, E. The science, technology, and applications of magnesium. *JOM* **1998**, *50*, 30–34. [CrossRef]
7. Tan, J.; Ramakrishna, S. Applications of magnesium and its alloys: A review. *Appl. Sci.* **2021**, *11*, 6861. [CrossRef]
8. Wang, S.; Zha, M.; Jia, H.; Yang, Y.; Wang, D.; Wang, C.; Gao, Y.; Wang, H. A review of superplastic magnesium alloys: Focusing on alloying strategy, grain structure control and deformation mechanisms. *J. Mater. Sci. Technol.* **2025**, *211*, 303–319. [CrossRef]
9. Luo, X.; Feng, Z.; Yu, T.; Luo, J.; Huang, T.; Wu, G.; Hansen, N.; Huang, X. Transitions in mechanical behavior and in deformation mechanisms enhance the strength and ductility of Mg-3Gd. *Acta Mater.* **2020**, *183*, 398–407. [CrossRef]
10. Wang, J.; Wang, X.; Yu, K.; Rupert, T.J.; Mahajan, S.; Lavernia, E.J.; Schoenung, J.M.; Beyerlein, I.J. Manipulating deformation mechanisms with Y alloying of Mg. *Mater. Sci. Eng. A* **2021**, *817*, 141373. [CrossRef]
11. Mo, N.; McCarroll, I.; Tan, Q.; Ceguerra, A.; Cairney, J.; Dieringa, H.; Huang, Y.; Jiang, B.; Pan, F.; Bermingham, M.; et al. Roles of Nd and Mn in a new creep-resistant magnesium alloy. *Mater. Sci. Eng. A* **2020**, *779*, 139152. [CrossRef]
12. Wang, J.; Xu, G.; Zeng, X.; Llorca, J.; Cui, Y. Exploring the correlation between solvent diffusion and creep resistance of Mg–Ga HCP alloys from high throughput liquid-solid diffusion couple. *Mater. Des.* **2021**, *197*, 109243. [CrossRef]
13. Zhang, L.; Yuan, Q.; Tan, J.; Dong, Q.; Lv, H.; Wang, F.; Tang, A.; Eckert, J.; Pan, F. Enhancing the room-temperature plasticity of magnesium alloys: Mechanisms and strategies. *J. Magnes. Alloys* **2024**, *12*, 4741–4767. [CrossRef]
14. Zhang, L.; Wu, M.; Xu, C.; Guo, S.; Ning, Z.; Cao, F.; Huang, Y.; Sun, J.; Yi, J. Creep behaviors of a Mg–Li based alloy at elevated temperatures. *Mater. Sci. Eng. A* **2021**, *827*, 142063. [CrossRef]
15. Kang, Y.; Pelton, A.; Chartrand, P.; Fuerst, C. Critical evaluation and thermodynamic optimization of the Al–Ce, Al–Y, Al–Sc and Mg–Sc binary systems. *Calphad* **2008**, *32*, 413–422. [CrossRef]
16. Ogawa, Y.; Ando, D.; Sutou, Y.; Somekawa, H.; Koike, J. Martensitic transformation in a β-type Mg–Sc alloy. *Shap. Mem. Superelast.* **2018**, *4*, 167–173. [CrossRef]
17. Liu, J.; Lin, Y.; Bian, D.; Wang, M.; Lin, Z.; Chu, X.; Li, W.; Liu, Y.; Shen, Z.; Liu, Y.; et al. In vitro and in vivo studies of Mg-30Sc alloys with different phase structure for potential usage within bone. *Acta Biomater.* **2019**, *98*, 50–66. [CrossRef] [PubMed]
18. Liu, C.; Ji, Y.; Tang, J.; Otsuka, K.; Wang, Y.; Hou, M.; Hao, Y.; Ren, S.; Luo, P.; Ma, T.; et al. A lightweight strain glass alloy showing nearly temperature-independent low modulus and high strength. *Nat. Mater.* **2022**, *21*, 1003–1007. [CrossRef] [PubMed]
19. Silva, C.; Kula, A.; Mishra, R.; Niewczas, M. The effect of Sc on plastic deformation of Mg–Sc binary alloys under tension. *J. Alloys Compd.* **2018**, *761*, 58–70. [CrossRef]
20. Ogawa, Y.; Ando, D.; Sutou, Y.; Koike, J. A lightweight shape-memory magnesium alloy. *Science* **2016**, *353*, 368–370. [CrossRef]
21. Yamagishi, K.; Ogawa, Y.; Ando, D.; Sutou, Y.; Koike, J. Room temperature superelasticity in a lightweight shape memory Mg alloy. *Scr. Mater.* **2019**, *168*, 114–118. [CrossRef]
22. Ando, D.; Ogawa, Y.; Suzuki, T.; Sutou, Y.; Koike, J. Age-hardening effect by phase transformation of high Sc containing Mg alloy. *Mater. Lett.* **2015**, *161*, 5–8. [CrossRef]
23. Ogawa, Y.; Ando, D.; Sutou, Y.; Koike, J. Aging effect of Mg–Sc alloy with α+β two phase microstructure. *Mater. Trans.* **2016**, *57*, 1119–1123. [CrossRef]
24. Ogawa, Y.; Sutou, Y.; Ando, D.; Koike, J. Aging precipitation kinetics of Mg–Sc alloy with bcc+hcp two-phase. *J. Alloys Compd.* **2018**, *747*, 854–860. [CrossRef]
25. Ogawa, Y.; Ando, D.; Sutou, Y.; Yoshimi, K.; Koike, J. Determination of α/β phase boundaries and mechanical characterization of Mg–Sc binary alloys. *Mater. Sci. Eng. A* **2016**, *670*, 335–341. [CrossRef]
26. Buch, F.; Mordike, B. Microstructure, mechanical properties and creep resistance of binary and more complex magnesium scandium alloys. In *Magnesium Alloys and Their Applications*; Kainer, K.U., Ed.; Wiley-VCH: Weinheim, Germany, 1998; pp. 145–150.
27. Mordike, B. Development of highly creep resistant magnesium alloys. *J. Mater. Process. Technol.* **2001**, *117*, 391–394. [CrossRef]
28. Mordike, B.; Ebert, T. Magnesium: Properties, applications and potential. *Mater. Sci. Eng. A* **2001**, *302*, 37–45. [CrossRef]
29. Mordike, B. Creeperesistant magnesium alloys. *Mater. Sci. Eng. A* **2002**, *324*, 103–112. [CrossRef]
30. Mordike, B.; Stulikova, I.; Smola, B. Mechanisms of creep deformation in Mg–Sc based alloys. *Metall. Mater. Trans. A* **2005**, *36*, 1729–1736. [CrossRef]
31. Durst, K.; Maier, V. Dynamic nanoindentation testing for studying thermally activated processes from single to nanocrystalline metals. *Curr. Opin. Solid State Mater. Sci.* **2015**, *19*, 340–353. [CrossRef]
32. Vagarali, S.S.; Langdon, T.G. Deformation mechanisms in hcp metals at elevated temperatures—I. Creep behavior of magnesium. *Acta Metall.* **1981**, *29*, 1969–1982. [CrossRef]
33. Staroselsky, A.; Anand, L. A constitutive model for hcp materials deforming by slip and twinning: Application to magnesium alloy AZ31B. *Int. J. Plast.* **2003**, *19*, 1843–1864. [CrossRef]

34. Ahmad, M.; Sosa, J.C.; Rongong, J. Characterisation of creep behaviour using the power law model in copper alloy. *J. Mech. Eng. Sci.* **2017**, *11*, 2503–2510. [CrossRef]
35. Mao, J.; Liu, W.; Li, D.; Zhang, C.; Ma, Y. The Strain Rate Sensitivity and Creep Behavior for the Tripler Plane of Potassium Dihydrogen Phosphate Crystal by Nanoindentation. *Micromachines* **2021**, *12*, 369. [CrossRef]
36. Viespoli, L.M.; Johanson, A.; Alvaro, A.; Nyhus, B.; Sommacal, A.; Berto, F. Tensile characterization of a lead alloy: Creep induced strain rate sensitivity. *Mater. Sci. Eng. A* **2019**, *744*, 365–375. [CrossRef]
37. Alkorta, J.; Martínez-Esnaola, J.M.; Sevillano, J.G. Critical examination of strain-rate sensitivity measurement by nanoindentation methods: Application to severely deformed niobium. *Acta Mater.* **2008**, *56*, 884–893. [CrossRef]
38. Somekawa, H.; Mukai, T. Nanoindentation creep behavior of grain boundary in pure magnesium. *Philos. Mag. Lett.* **2010**, *90*, 883–890. [CrossRef]
39. Han, L.; Northwood, D.O.; Nie, X.; Hu, H. The effect of cooling rates on the refinement of microstructure and the nanoscale indentation creep behavior of Mg–Al–Ca alloy. *Mater. Sci. Eng. A* **2009**, *512*, 58–66. [CrossRef]
40. Hu, J.; Zhang, W.; Bi, G.; Lu, J.; Huo, W.; Zhang, Y. Nanoindentation creep behavior of coarse-grained and ultrafine-grained pure magnesium and AZ31 alloy. *Mater. Sci. Eng. A* **2017**, *698*, 348–355. [CrossRef]
41. Fang, L.; Wang, J.; Xu, C.; He, X.; Xu, G.; Tao, X.; Ouyang, Y.; Du, Y. Interdiffusion behaviors and mechanical properties in Zr–Nb–Hf system. *Calphad* **2023**, *83*, 102634. [CrossRef]
42. Haußmann, L.; Neumeier, S.; Bresler, J.; Keim, S.; Pyczak, F.; Göken, M. Influence of Nb, Ta and Zr on the Interdiffusion Coefficients and Solid Solution Strengthening of γ-TiAl Single Phase Alloys. *Metals* **2022**, *12*, 752. [CrossRef]
43. Mao, S.; Wang, C.; Li, N.; Wang, J.; Chen, Y.; Xu, G.; Guo, Y.; Cui, Y. Kinetic diffusion multiple: A high-throughput approach to screening the composition-microstructure-micromechanical properties relationships. *Calphad* **2018**, *61*, 219–226. [CrossRef]
44. Conrad, H. Thermally activated deformation of metals. *JOM* **1964**, *16*, 582–588. [CrossRef]
45. Dong, R.; Xu, G.; Ko, W.; Wang, J.; Tao, X.; Cui, Y. Diffusion behaviors and atomic mobilities in Mg–Sc hcp and bcc alloys: Investigation via single-phase and multi-phase diffusion couples. *Calphad* **2021**, *72*, 102228. [CrossRef]
46. Yu, Y.; Zhou, Z.; Tao, X.; Wang, J.; Xu, G.; Cui, Y. Measurement of the diffusion coefficient in Mg–Sn and Mg–Sc binary alloys. *Int. J. Mater. Res.* **2022**, *113*, 391–399. [CrossRef]
47. Li, N.; Li, Y.; Yang, L.; Wang, C. High-throughput investigation of temperature-dependent creep mechanisms in Mg–Zn alloys using nanoindentation. *Phys. Scr.* **2025**, *100*, 055933. [CrossRef]
48. Choi, I.-C.; Yoo, B.-G.; Kim, Y.-J.; Jang, J. Indentation creep revisited. *J. Mater. Res.* **2012**, *27*, 3–11. [CrossRef]
49. Zhao, J.; Jin, Z.; Huang, P. Diffusion quadruples for the determination of quaternary phase diagrams applied to FeCoNiCr system. *Scr. Metall.* **1988**, *22*, 1825–1829. [CrossRef]
50. Cheng, K.; Xu, H.; Ma, B.; Zhou, J.; Tang, S.; Liu, Y.; Sun, C.; Wang, N.; Wang, M.; Zhang, L.; et al. An in-situ study on the diffusion growth of intermetallic compounds in the Al–Mg diffusion couple. *J. Alloys Compd.* **2019**, *810*, 151878. [CrossRef]
51. Long, X.; Shen, Z.; Jia, Q.; Li, J.; Dong, R.; Su, Y.; Yang, X.; Zhou, K. Determine the unique constitutive properties of elastoplastic materials from their plastic zone volution under nanoindentation. *Mech. Mater.* **2022**, *175*, 104485. [CrossRef]
52. Robertson, C.; Poissonnet, S.; Boulanger, L. Plasticity in ion-irradiated austenitic stainless steels. *J. Mater. Res.* **1998**, *13*, 2123–2131. [CrossRef]
53. Chiang, S.; Marshall, D.; Evans, A. The response of solids to elastic/plastic indentation. *Proc. R. Soc. Lond.* **2013**, *53*, 298–311.
54. Baker, S.; Liu, J. *Nanoindentation Techniques, Reference Module in Materials Science and Materials Engineering*; Elsevier: Amsterdam, The Netherlands, 2016.
55. Zhang, Z.; Li, Y.; Wu, W. Effects of loading rate and peak load on nanoindentation creep behavior of DD407Ni-base single crystal superalloy. *Trans. Nonferrous Met. Soc. China* **2022**, *32*, 206–216. [CrossRef]
56. Wang, Y.; Liu, W.; Ma, Y.; Huang, Y.; Tang, Y.; Cheng, F.; Yu, Q. Indentation size effect and micromechanics characterization of intermetallic compounds in the Au–Sn system. *Mater. Sci. Eng. A* **2014**, *610*, 161–170. [CrossRef]
57. Wang, J.; Zhu, G.; Wang, D.; Liu, Y.; Zeng, X.; Knezevic, M. Creep strength of Mg–Y–Zn alloys containing long-period stacking ordered structures. *J. Alloys Compd.* **2025**, *1024*, 180110. [CrossRef]
58. Sun, Y.; Huo, Y.; Yu, W.; Yan, Z.; Wang, Z.; Li, Z.; Wang, Z.; Chen, H.; Jiang, A.; Wang, X. Microstructure and nanoindentation creep behavior of NiAlCrFeMo high-entropy alloy. *J. Alloys Compd.* **2025**, *1020*, 179330. [CrossRef]
59. Wu, S.; Zhang, J.; Zhang, Z.; Xu, C.; Nie, K.; Niu, X. A high strength and good ductility Mg–Y–NI–TI alloy with long period stacking ordered structure processed by hot rolling and aging treatment. *Mater. Sci. Eng. A* **2015**, *648*, 134–139. [CrossRef]
60. Zhang, R.; Zhao, S.; Ding, J.; Chong, Y.; Jia, T.; Ophus, C.; Asta, M.; Ritchie, R.; Minor, A. Short-range order and its impact on the CrCoNi medium-entropy alloy. *Nature* **2020**, *581*, 283–287. [CrossRef] [PubMed]

Disclaimer/Publisher's Note: The statements, opinions and data contained in all publications are solely those of the individual author(s) and contributor(s) and not of MDPI and/or the editor(s). MDPI and/or the editor(s) disclaim responsibility for any injury to people or property resulting from any ideas, methods, instructions or products referred to in the content.

materials

MDPI

Article

Environmental Factors Influencing Stress Corrosion Cracking Behavior of Austenitic Stainless Steels in Simulated Seawater

Chun-Ping Yeh *, Kun-Chao Tsai and Jiunn-Yuan Huang

National Atomic Research Institute (NARI), 1000 Wenhua Rd., Longtan District, Taoyuan City 32546, Taiwan
* Correspondence: cpyeh@nari.org.tw; Tel.: +886-3-471-1400

Abstract

Grade 304L stainless steel canisters are susceptible to crevice corrosion in marine environments. In the present study, white emery was utilized to create a simulation of dust accumulation. The corrosion testing was conducted at two distinct temperatures (35 °C and 45 °C) and three levels of relative humidity (45%, 55%, and 70% relative humidity). The chloride deposition density levels tested were 0.1 g/m^2 and 1 g/m^2. The test durations were 8000 h and 23,000 h. It is evident that with a chloride deposition density of 0.1 g/m^2 at a temperature of 45 °C and a relative humidity of 70%, the onset of stress corrosion cracking (SCC) occurred after 8000 h in the white emery deposition tests. In contrast, at a 1 g/m^2 chloride deposition density, the polytetrafluoroethylene (PTFE) crevice former test specimen exhibited continuous transgranular SCC within the same period. These quantitative findings emphasize the critical roles of salt load and environmental severity in the initiation of SCC.

Keywords: marine environment; chloride concentration; stainless steel; SCC; dust

1. Introduction

It is critical to ensure the durability of dry storage canisters to ensure the safe and secure management of spent nuclear fuel (SNF), particularly in facilities located near coastal regions. Austenitic stainless steels (ASS), particularly the 304L and 316L grades, are frequently utilized in the fabrication of these canisters due to their proven superiority in terms of resistance to corrosion [1]. However, these materials are prone to chloride-induced SCC (CISCC), a significant degradation mechanism that is further exacerbated by salt contamination from airborne particles in marine environments [2,3]. The process commences with the dissipation of heat from the decaying SNF, which reduces the surface temperature of the canisters to below 100 °C. This initiates the deliquescence of the salts that have been deposited, resulting in the production of a chloride-rich aqueous layer that promotes localized corrosion [4,5]. In order to guarantee the durability and safety of these storage systems, it is vital to gain a comprehensive understanding of the contributing environmental factors that drive CISCC.

Environmental factors, including chloride concentration, relative humidity (RH), and temperature, perform a pivotal function in the progression of CISCC. It has been demonstrated that elevated temperatures enhance corrosion kinetics and diffusion, while a high RH sustains a thin electrolyte layer, intensifying anodic dissolution and crack propagation. The Nuclear Regulatory Commission (NRC) has identified CISCC as a pivotal challenge threatening the safe long-term operation of dry storage systems for SNF [6].

Numerous researchers have studied the corrosive deterioration of stainless steel in marine surroundings. Kim et al. investigated the CISCC susceptibility of 0.6–1 M $NaCl_2$ deposited on 304L–ER308L welded plates at 50–60 °C with 20–30% RH. The results demonstrated that lowering the temperature from 60 °C to 50 °C did not prevent CISCC. Conversely, a reduction in humidity to 20% RH effectively inhibited the formation of CISCC. This suggests that the inhibition of NaCl deliquescence under conditions of reduced humidity was insufficient to promote CISCC [3]. Zhao et al. conducted an analysis of the corrosion behavior of 2205 duplex SS under conditions involving the U-bending deformation and the application of $MgCl_2$ droplets subjected to wet–dry cycling [7]. Dhaiveegan et al. conducted a three-year investigation into the corrosion behavior of 304 and 316L stainless steels when exposed to the Indian marine environment in the Chennai region. Based on their report, 304 SS does not demonstrate behaviors associated with passivation, whereas 316L SS reveals such behaviors in the marine surroundings of Chennai, characterized by high relative humidity [8]. Burkert et al. reported that stainless steel specimens with crevices exhibited considerably deeper pits when exposed to a marine environment [9]. Lv et al. reported that the initiation of pits is related to the dissolution of MnS inclusions [10]. In a simulated marine environment, the maximum pit depth of 304 stainless steel increased with corrosion time [10].

Austenitic stainless steel canister materials are susceptible to CISCC when exposed to chloride-containing sea salts due to the formation of a film of deliquescent salts on the canister surface that can retain moisture [11,12]. Deliquescent sea salts contain chloride ions that are highly aggressive. They induce the degradation of the passive film and the initiation of SCC. A hypothesis has been proposed that links the crevice corrosion characteristics of stainless steel to the process of passive film development within crevices [13]. As the chloride concentration increases, the stability of the passive film decreases, leading to more severe crevice corrosion phenomena [14]. It has been demonstrated that the canisters undergo a gradual accumulation of dust on their surfaces with the presence of chloride salts over time. The deposition of dust on the canister surface with the presence of chloride salts establishes an ambient environment favorable for water accumulation and subsequent localized corrosion, such as crevice corrosion.

Crevice corrosion has been shown to enhance CISCC by means of trapping deposits of chloride on the surface of the canister. Crevice corrosion initiation is affected by aggressive Cl^- ions, the characteristics of the crevice, and the ambient temperature. The characteristics of the crevice are established in that the canister is in physical contact with the storage module's structural support, a feature present in both vertical and horizontal canister configurations [15]. The behavior of crevice corrosion can be explained by means of the mechanism of critical crevice solution (CCS). The depletion of oxygen in the crevice is believed to be the underlying cause of the acidification of the solution, the subsequent breakdown of the passive film, and the initiation of crevice corrosion behavior. It has been demonstrated that this mechanism is the result of the effect of aggressive ions, such as chloride ions. Accumulation of these ions in the crevice is followed by depassivation, resulting from the active dissolution of the base metal [16,17]. The susceptibility of austenitic stainless steels to SCC has been well documented, particularly in aggressive surroundings, for example, Cl^- ions, which can result in CISCC [18–20]. The increased volumetric presence of corrosion products at crack sites could create localized stresses that enhance the initiation of stress corrosion cracking. Mayuzumi et al. revealed that stress corrosion cracking initiation occurs at the bottom of crevice corrosion areas in the presence of chloride ions [21]. Tani and colleagues suggested that crevice corrosion leads to SCC initiation on surfaces subjected to saline particles [22].

The chloride deposition density on canister surfaces can be estimated by analyzing the sea salt deposit's areal density and the local environmental conditions. Guo et al. reported on the corrosion behavior of 304 stainless steel in the presence of atmospheric conditions involving a mixture of sodium chloride (NaCl) and magnesium chloride ($MgCl_2$) at a temperature of 21 °C and a relative humidity of 45%. It was determined that pits of a bowl-shaped configuration form in solutions that contain a mixture of salts, while crevice corrosion occurs beneath NaCl crystals [23]. Dong et al. examined the stress corrosion cracking susceptibility of $MgCl_2$, depositing it on 316L SS at 75 °C and 70% RH. It was determined that transgranular SCC (TGSCC) occurred at pit edges in the welded region [24]. The influence of salt on CISCC in SS304L at 90 °C and 70% RH was examined by Scatigno et al. At a 1.7 g/m^2 chloride deposition density of $MgCl_2$, CISCC was found in the environment [25]. Shoji and colleagues investigated the effect of various environmental factors on the initiating behavior of CISCC. Specifically, they examined how relative humidity, temperature, and chloride deposition density influenced the initiation process. They concluded that the presence of $MgCl_2$, a constituent of sea salt, has been demonstrated to promote the occurrence of low-temperature CISCC in SS304L and SS316L [26], and their works were summarized by Scatigno et al. [27]. Cook et al. provided a comprehensive summary of the ranges of relative humidity where CISCC occurred, along with the chloride deposition density of $MgCl_2$ and seawater at 25 g/m^2, based on the findings of Shoji [28]. Prosek et al. examined the CISCC phenomenon in SS304 and SS316L at a 260 g/m^2 chloride concentration, which was achieved by the deposition of $MgCl_2$ droplets [29]. In addition, based on the work of Ornek et al., the chloride deposition density at the crack tips of SS304L was found to be saturated during salt loading tests. Accordingly, the level of surface chloride did not have a significant impact on the rate of crack propagation [30].

The Nuclear Regulatory Commission (NRC) conducted experiments using 304 stainless steel U-bend specimens at temperatures ranging from 27 °C to 60 °C. According to the findings of the NRC, the chloride deposition density threshold for initiating stress corrosion cracking was determined to be 0.1 g/m^2 [31]. Furthermore, the chloride deposition density measured on surfaces of the canister exposed to sea salt was found to be merely 1 g/m^2, which is considerably lower than the expected value over a 19-year period [32]. The Central Research Institute of Electric Power Industry (CRIEPI) subjected 304L stainless steel samples to a 50 °C and 35% RH testing environment for over 2000 h. CRIEPI revealed that the chloride deposition density threshold for SCC initiation was 0.8 g/m^2 [33]. Furthermore, the canister surface temperature was 40–52 °C [26], and the inlet temperature of the canister was 35 °C where surveillance coupons were located [34].

Qiao et al. demonstrated the SCC phenomenon in 321 SS when exposed to a $MgCl_2$ solution. The interaction between primary cracks and discontinuous microcracks has been found to increase the effective stress intensity factor. Consequently, this interaction promotes crack coalescence, thereby resulting in the mechanical rupture of the ligaments that exist between the aforementioned cracks in stainless steel [35]. Masuda et al. examined the stress corrosion cracking behavior of 304 stainless steel, focusing on its slip deformation, pit growth, and surface potential distribution. SCC testing was executed at temperatures of 70 °C and relative humidity levels of 28% utilizing $MgCl_2$ droplets. Usually, discrete cracks were detected in the region proximate to the crack tip [36].

The accumulation of dust can lead to a notable increase in surface roughness. This phenomenon is associated with the formation of more profound grooves on the surface of the material, where a high accumulation of dust and ion concentrations is observed. Consequently, this can result in the initiation of corrosion at relatively early stages [37]. In addition, an elevated roughness promotes the interaction between the corrosive agents present in moisture and the metal surface.

Despite the extensive research conducted on CISCC under atmospheric chloride de posits, there is a lack of studies employing a synthetic, high-purity particulate analogous to dust-particle effects. The existing literature is deficient in its lack of a reproducible surrogate with well-controlled size, shape, and mass. These elements are critical for the systematic evaluation of particulate-driven crevice corrosion phenomena. In this study, white emery was utilized as a model dust layer. Therefore, this study proposes a novel approach that involves the integration of a controlled white emery surrogate with well-defined environmental factors. The objective of this study is to address the critical knowledge gap concerning dust–salt particulate-driven CISCC in nuclear-grade austenitic stainless steel canisters.

2. Experimental Procedure

Figure 1 presents a composite image of the sample and the crevice former. To simulate the behavior of crevice corrosion at the location where the canister contacts the storage module support structure, we used specimens deposited with salt and wrapped with a crevice former to create a crevice configuration. The specimen's thickness was precisely 3 mm. The chemical composition of the 304L SS utilized for this study is provided in Table 1. The sample was composed of 304L SS, and the crevice former was composed of polytetrafluoroethylene. Prior to the execution of the crevice corrosion examination, the surface of the specimen was subjected to a grinding process using 2000-grit abrasive paper.

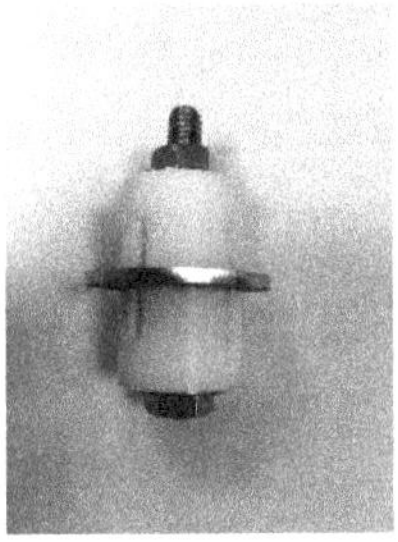

Figure 1. Assembled image of the 304L stainless steel specimen and crevice former utilized for the crevice corrosion test.

Table 1. Chemical composition of the 304L stainless steel utilized in this study.

Element	C	Si	S	Cr	Ni	Mn	Fe
Wt%	0.017	0.450	0.029	18.000	9.000	1.540	Bal.

The present study employed white emery as a means to simulate the accumulation of dust on the as-machined specimens during the testing of crevice corrosion induced by dust deposition on the canister surface. Figure 1 demonstrates the dimensions of the as-machined specimens, while the surface roughness was 4.9 μm. The as-machined samples were prepared from 304L SS, while the white emery was prepared from alumina. Each experimental condition was replicated.

Table 2 illustrates the chemical composition of the se salt utilized in the present research, which adheres to the standards outlined in ASTM D 1141-98 (2013) [38], as indicated under Formula a, Table X1.1. In the standard crevice corrosion test, the specimens were exposed to a salt spray using experimental equipment and synthetic seawater at a concentration of 3.5 wt%. Subsequently, the specimens were dried on a hot plate at 65 °C for a duration of 15 min. Each specimen's mass difference was evaluated to determine the salt load in g/m^2 after drying. The sample chloride deposition density was maintained at 0.1 g/m^2 and 1 g/m^2. The sample and the crevice former were affixed to each other with an M6 bolt, and torque was applied to 1.13 Nm to create the crevice configuration. Similarly, this research used white emery to imitate accumulated dust on the as-machined specimens' surfaces to simulate crevice corrosion originating from dust deposition. A total of 0.5 g of white emery was uniformly dispersed via spray application around the circumference of the sample, rather than being fixed with a crevice former. Figure 2 [39] presents a schematic representation of the specimen after exposure to artificial seawater and white emery. All exposures were conducted within temperature- and humidity-controlled chambers, where the temperature was maintained at 35 °C or 45 °C and the relative humidity (RH) was set at 45%, 55%, or 70%. The duration of the exposures was 8000 h or 23,000 h, as detailed in Table 3. Metallographic procedures were utilized to prepare the specimens for microstructural analysis. The preparation of the samples included an ultrasonic cleansing step with deionized water, subsequent drying, and mounting in resin. Aluminum oxide powder was utilized in the polishing of the mounted specimens, which were then subjected to thorough examinations aimed at analyzing their microstructure and surface characteristics. In addition, a JSM-7100F field-emission SEM (JEOL, Tokyo, Japan) equipped with an EBSD detector (Oxford Instruments, Abingdon, UK) was utilized to investigate the SCC characteristics of the tested specimens. The kernel average misorientation (KAM) map and the strain countering map for the samples can be extracted by entering the EBSD map into the HKL-Channel 5 software (5.12.67.0, Oxford Instruments, Abingdon, UK) for data processing.

Table 2. Chemical composition of the sea salt used in this work.

Composition	$MgCl_2$	NaCl	Na_2SO_4	KCl	$CaCl_2$	KBr	$NaHCO_3$	$SrCl_2$	NaF	H_3BO_3
wt%	26.460	58.490	9.750	1.645	2.765	0.238	0.477	0.095	0.007	0.071

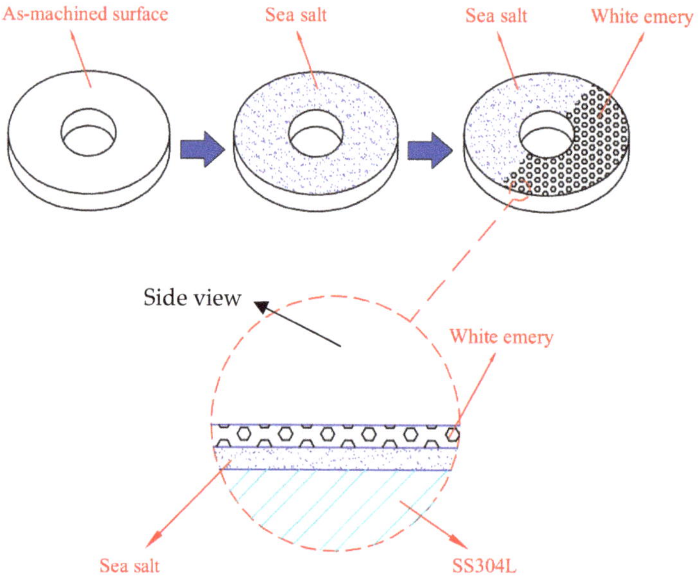

Figure 2. Schematic representation of the specimen sprayed with white emery and synthetic seawater [39].

Table 3. Test conditions of the specimens.

Parameter	Unit	Value
Temperature	°C	35, 45
Relative humidity	%	45, 55, 70
Chloride deposition density	g/m^2	0.1, 1
Test duration	Hours	8000, 23,000
Testing method for sample surface	N/A	White emery deposition, crevice former

3. Results

3.1. Surface Morphology Examination

As demonstrated in Figure 3, the 304L specimen that was sprayed with 0.5 g of white emery and exposed to 1 g/m^2 chloride at a temperature of 45 °C and a relative humidity of 70% reveals the presence of a corrosion band along the semicircular deposit. The corrosion products form a belt approximately 1 mm in width at the deposit edge, indicating localized crevice corrosion beneath the white emery layer. Figures 4 and 5 illustrate the SEM images of the specimens that were subjected to the white emery spray and a 0.1 g/m^2 chloride concentration. The SEM micrographs demonstrate that, under conditions of a lower salt load (0.1 g/m^2), there is a marked increase in the corrosion severity from 35 °C to 45 °C at 70% RH. Furthermore, the initial appearance of rust spots is observed at a relative humidity of at least 55%, and their progression to interconnected "tunnels" is evident only at 70% RH (see Figure 5d). Notably, at a temperature of 35 °C, even after 23,000 h of exposure, only shallow pits are observed, suggesting a different susceptibility to rust under these conditions.

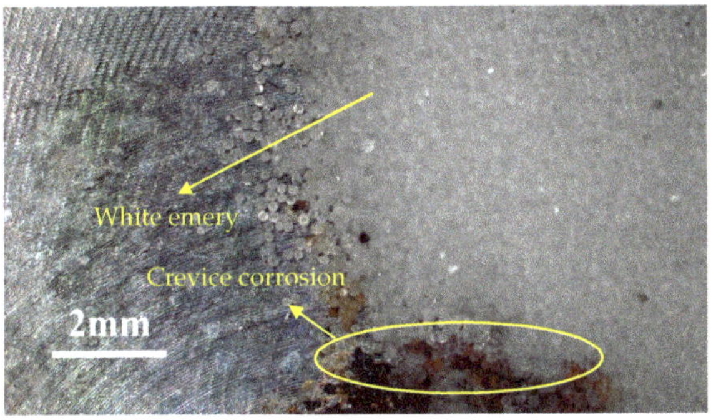

Figure 3. Macrographs of specimens subjected to white emery spray and deposited at a 1 g/m^2 chloride concentration at a temperature of 45 °C and a relative humidity of 70%.

Figures 6 and 7 are the macrographs of the 304L stainless steel specimens sprayed with white emery and wrapped with a crevice former at two different temperature levels; all specimens were deposited at a chloride concentration of 1 g/m^2. Figures 6a and 7a demonstrate the presence of crevice corrosion on the specimens subjected to white emery spray and enveloped within the crevice former under conditions of 35 °C. The specimens wrapped with a crevice former, as shown in Figure 7a, exhibit larger corrosion pits than those sprayed with white emery, shown in Figure 6a. Significant corrosion of the samples is revealed in Figure 7b, with products of corrosion on exposed surfaces. The specimen's appearance indicates a decrease in brightness, with a surface color that appears yellowish-brown. Furthermore, the areas of rust on the specimens subjected to white emery spray and wrapped with a crevice former increased as the temperature increased, as shown in Figures 6 and 7. It has been demonstrated that exposure to constant humidity, in conjunction with rising temperatures, results in accelerated diffusion and electrochemical kinetics, which, in turn, gives rise to an enhancement in corrosion rates [33,40]. Moreover, it appears that the specimens enveloped within the crevice former, Figure 7a,b, are more corroded than those sprayed with white emery, Figure 6a,b, at both temperature levels.

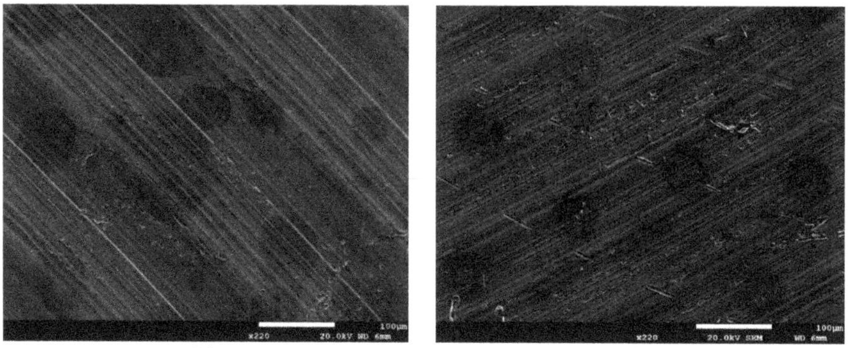

(**a**) (**b**)

Figure 4. SEM micrographs of specimens subjected to white emery spray and deposited at a 0.1 g/m^2 chloride concentration and 70% RH after 8000 h of testing at: (**a**) 35 °C, and (**b**) 45 °C.

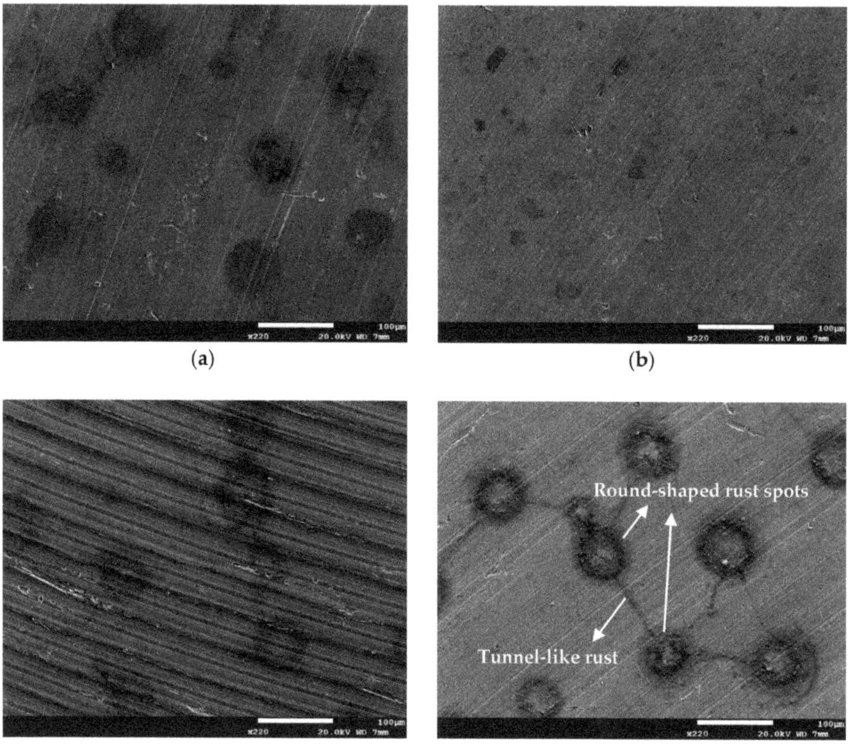

(**a**) (**b**)

(**c**) (**d**)

Figure 5. SEM micrographs of specimens subjected to white emery spray and deposited at a 0.1 g/m^2 chloride concentration after 23,000 h of testing at: (**a**) 35 °C and 70%RH, (**b**) 45 °C and 45% RH, (**c**) 45 °C and 55% RH, and (**d**) 45 °C and 70% RH.

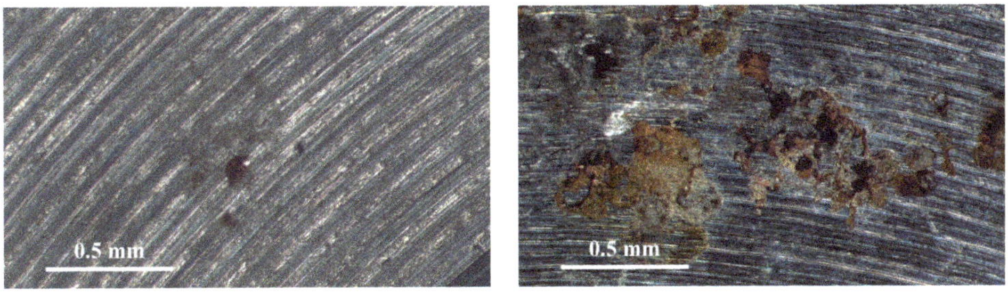

(**a**) (**b**)

Figure 6. Macrographs of specimens subjected to white emery spray and deposited at a 1 g/m^2 chloride deposition density and 70% RH after 8000 h of testing at: (**a**) 35 °C, and (**b**) 45 °C.

(**a**) (**b**)

Figure 7. Macrographs of specimens enveloped within the crevice former and deposited at a 1 g/m^2 chloride concentration and 70% relative humidity after 8000 h of testing at: (**a**) 35 °C, and (**b**) 45 °C.

3.2. Investigation of EDS and EBSD

Figure 8 demonstrates the EDS examinations of the corrosion pits in adjacent cracks on the specimen wrapped with a crevice former and deposited with a 1 g/m^2 chloride deposition density at 35 °C and 70% RH. The findings of the EDS analysis of the corrosion pits are presented in Table 4. At point B, the iron content was significantly higher and the oxygen content was lower in comparison to points C and E. This observation suggests that the oxide product was peeled off, thereby unveiling the metal surface. Furthermore, the chlorine and sulfur content of point A, the matrix of stainless steel, was not detectable due to the extremely low concentration levels. Trace quantities of Cl were observed at points B, C, D, and E, which were located in the corrosion products. It was determined that points B, C, D, and E contained higher concentrations of sulfur in comparison to point A, probably due to the sulfates included in the synthesized seawater.

Table 4. EDS results on the selected location of Figure 8 in weight percentage (wt%).

Location	Fe	Cr	Ni	Mn	O	S	Cl
A	69.9	18.9	8.00	1.80	1.40	0.00	0.00
B	23.6	46.5	3.60	1.50	23.0	1.50	0.30
C	14.1	35.6	3.90	0.50	43.2	2.30	0.50
D	19.6	28.5	2.40	0.70	46.2	2.30	0.40
E	15.1	29.6	3.90	0.60	47.8	2.80	0.30

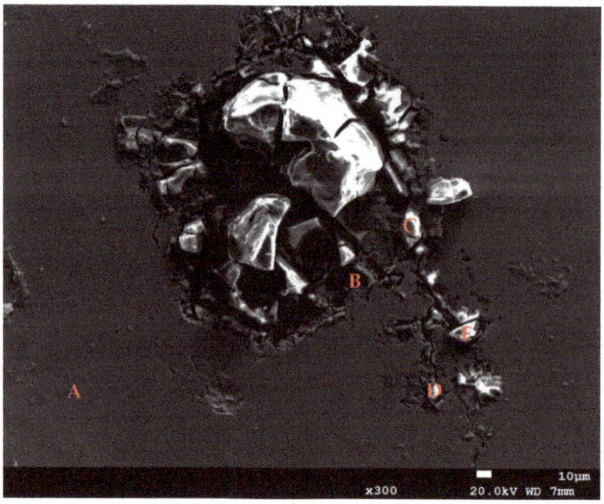

Figure 8. EDS examination of corrosion pits on the specimen enveloped within the crevice former and deposited at a 1 g/m^2 chloride concentration at 35 °C and 70% RH (EDS analysis results of points A–E are listed in Table 4).

Crevice corrosion initiation is significantly affected through the presence of chloride ions. The critical crevice solution (CCS) mechanism suggests that the onset of crevice corrosion is dependent on the accumulation of corrosive ions, particularly chloride (Cl^-), at the crevice sites. This phenomenon has been observed to result in the onset of localized corrosion, characterized by its notably aggressive nature. The severity of this localized corrosion has the capacity to exert a substantial detrimental effect on stainless steel's passive film [41]. Compared to points C and E, point B contains higher concentrations of iron but lower concentrations of oxygen. This suggests that the oxide product was stripped off, exposing the surface of the metal.

Figure 9 and Figure 10 respectively, show SEM micrographs that reveal the morphology of SCC on the specimens sprayed with white emery and deposited at a chloride deposition density of 0.1 g/m^2 and 70% RH at 45 °C for 8000 h and at 35 °C and 45 °C for 23,000 h. These findings indicate that SCC cracks initiate only under sufficiently aggressive conditions (70% RH). Following an exposure duration of 8000 h at a temperature of 45 °C, the formation of discontinuous microcracks becomes evident in the region adjacent to the rust spots. These microcracks exhibit a greater length when compared to those observed in specimens that were exposed for 23,000 h at a temperature of 35 °C. The extension of the 0.1 g/m^2 tests to 23,000 h at 45 °C results in the transformation of these microcracks into an almost continuous crack. This finding suggests that temperature exerts a more significant influence on crack morphology than exposure duration under low salt loading conditions.

Figures 11 and 12 are SEM micrographs showing the morphology of stress corrosion cracking on specimens subjected to white emery spray and wrapped with a crevice former at two different temperature levels. All of the specimens were deposited with a 1 g/m^2 chloride deposition density. Figure 11a shows discontinuous SCC cracks on the specimen tested at 35 °C, while Figure 11b exhibits distinct cracks of SCC on specimens exposed to 45 °C. Similarly, the sample exposed to 45 °C, as illustrated in Figure 9, exhibited discrete SCC cracks on the specimens deposited with a 0.1 g/m^2 chlorine deposition density, while Figure 11b demonstrates continuous cracks of SCC on the specimens deposited with a 1 g/m^2 chloride deposition density. Figures 9, 10a, and 11a illustrate short, shallow cracks on the specimens, in contrast to Figures 11b and 12a,b, which illustrate long, deep cracks on the specimens. According to the assumption, microcracks first nucleate non-sequentially at appropriate locations and under appropriate conditions, and the process of crack growth is characterized by the formation of microcracks, which subsequently coalesce through the rupture of ligaments that connect them, resulting in the formation of a primary crack [35,36].

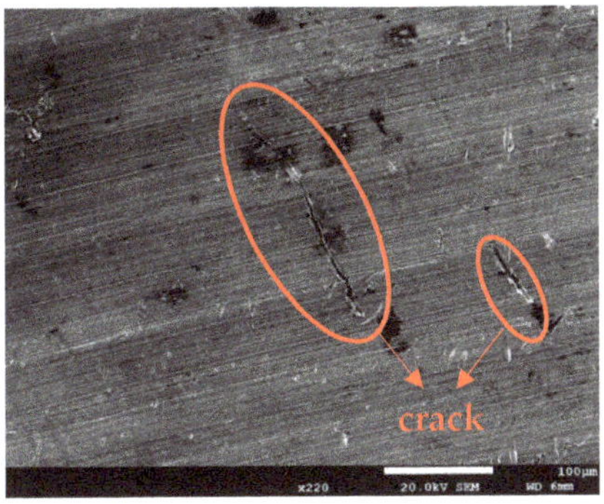

Figure 9. SEM morphologies of SCC on specimens subjected to white emery spray and deposited at a 0.1 g/m^2 chloride concentration after 8000 h of testing at a temperature of 45 °C and a relative humidity of 70%.

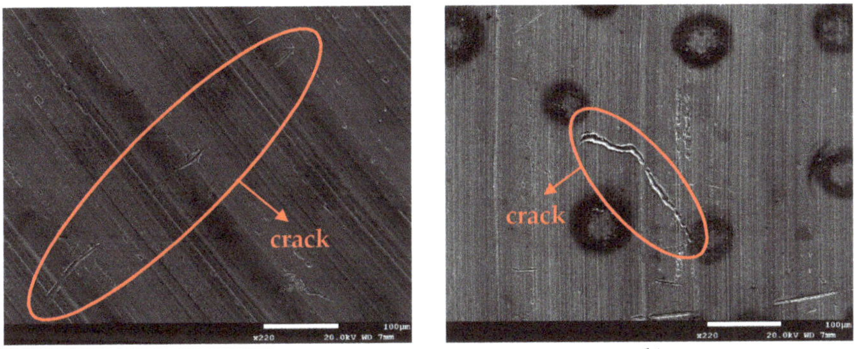

(**a**) (**b**)

Figure 10. SEM morphologies of SCC on specimens subjected to white emery spray and deposited at a 0.1 g/m^2 chloride deposition density and 70% RH after 23,000 h of testing at: (**a**) 35 °C, and (**b**) 45 °C.

(**a**) (**b**)

Figure 11. SEM morphologies of SCC on specimens subjected to white emery spray and deposited at a 1 g/m^2 chloride deposition density and 70% RH after 8000 h of testing at: (**a**) 35 °C, and (**b**) 45 °C.

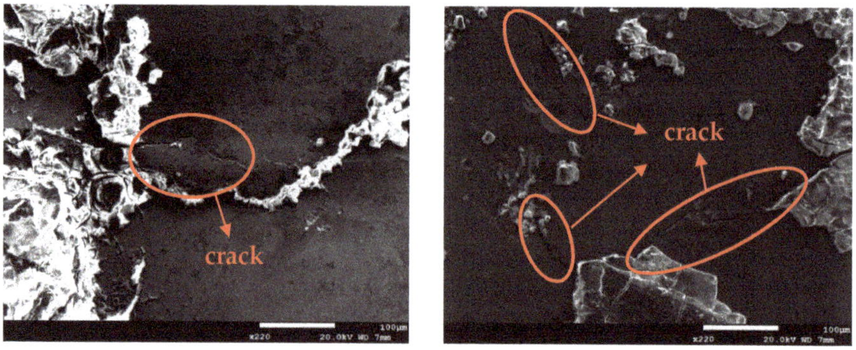

(**a**) (**b**)

Figure 12. SEM morphologies of SCC on specimens enveloped within the crevice former and deposited at a 1 g/m^2 chloride deposition density and 70% RH after 8000 h of testing at: (**a**) 35 °C, and (**b**) 45 °C.

Based on the aforementioned assumption, the specimens subjected to white emery spray and deposited with a 0.1 g/m^2 chloride deposition density and exposed to 45 °C, and deposited with a 1 g/m^2 chloride deposition density and exposed to 35 °C were subjected to 8000 h of testing. The specimens exhibited the presence of a few shallow and short cracks, which could be attributed to the limited number of chlorine ions transported to the sites of crevices in the specimens. Owing to the extremely low chloride concentration level of the specimens tested, it is imperative to establish an elevated-temperature environment to enhance the transportation of a sufficient quantity of Cl^- to the site of crevice formation, thereby facilitating cracking nucleation. The maps of chlorine acquired by EDS mapping, as demonstrated in Figure 13b, provide the supporting evidence for the above argument. This will be discussed in detail below. After testing for 8000 h, no cracks of SCC were investigated on the specimens subjected to white emery spray and tested at a temperature of 35 °C, whereas cracks were observed on the specimens tested at 45 °C, a higher-temperature environment, and on the specimens deposited with a chloride deposition density of 0.1 g/m^2, as shown in Figures 4 and 9. The occurrence of SCC cracks was observed at a temperature of 35 °C only when the test duration was increased to 23,000 h, as illustrated in Figure 10. The chloride deposition density of 0.1 g/m^2 was likely insufficient to initiate stress corrosion cracking on the specimens subjected to the decreased temperature and test duration. The increased volume of corrosion products likely generated a localized stress state beneath the white emery, thereby functioning similarly to a crevice former. This phenomenon is believed to be a contributing factor to the development of SCC. Moreover, an investigation of the crack length revealed a tendency for it to increase with temperature when comparing Figure 12a,b.

Figure 13 presents the EDS mapping for the crack regions of the specimens wrapped with a crevice former and deposited with a chloride deposition density of 1 g/m^2 at 35 °C and 70% RH. The crack region is rich in chlorine and oxygen, as illustrated in Figure 13b,c, while it is depleted of iron, as demonstrated in Figure 13d. The presence of corrosion products of iron in the vicinity of the crack region suggests that the process of corrosion resulted in a depletion of iron at the site of the crack. According to the critical crevice solution (CCS) model, the depletion of oxygen within the crevice can lead to the acidification of the solution at crevice sites. Consequently, breakdown of the passive film and initiation of crevice corrosion can occur. This model emphasizes the accumulation of aggressive ions (e.g., Cl^-) at crevice sites, followed by depassivation and subsequent base metal dissolution [17]. The enhanced chloride concentration has been demonstrated to lower local pH and acidify the solution. Consequently, the passive film is destabilized and localized crevice corrosion is initiated, thereby promoting crack formation.

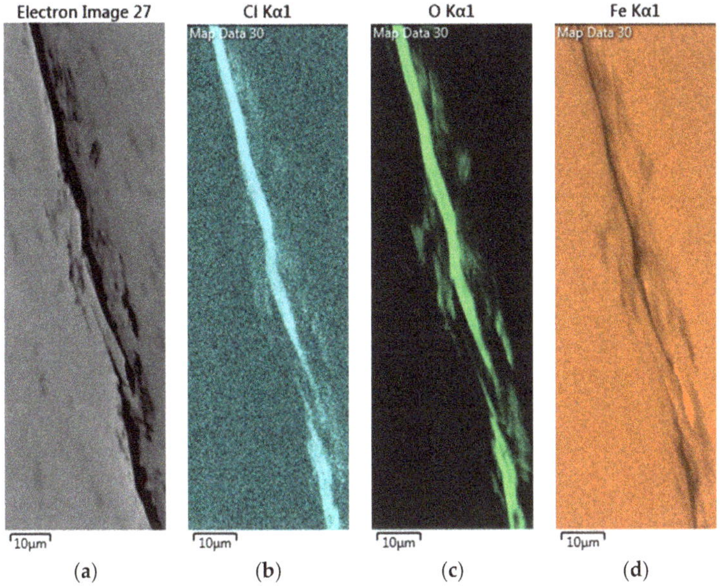

Figure 13. EDS mapping for the crack areas of a specimen enveloped within the crevice former and deposited with a chloride deposition density of 1 g/m^2 at 35 °C and 70% RH: (**a**) IQ (Image quality) figure, (**b**) Cl mapping, (**c**) O mapping, and (**d**) Fe mapping.

Figure 14 illustrates the EBSD and EDS maps for the crack areas of the specimen enveloped within the crevice former and deposited with a 1 g/m^2 chloride deposition density exposed to 35 °C and 70% relative humidity. Figure 14a,f demonstrate that the specimens were cracked by a TGSCC mode, which is perfectly consistent with previously reported observations of stainless steel induced by chlorides [22]. Figures 14b and 14c, respectively, indicate enriched manganese and sulfur at the top right and bottom left of this figure. The distribution of manganese and sulfur within the metal is related to the existence of non-metallic MnS inclusions. The strain countering map and kernel average misorientation (KAM) map (Figure 14d,e) illustrate that the stress is predominantly concentrated in the vicinity of the crack areas. In addition, as illustrated in Figure 14d, high plastic strain is observed in the regions adjacent to the cracks, suggesting a probable correlation with the plastic strain present at the tip of the cracks. The propagation of cracks was found to be associated with the degree of plastic strain present at the tips of the cracks. EBSD analyses confirm that SCC proceeds via a TGSCC mechanism, consistent with chloride-induced cracking in 304L SS [22]. The presence of MnS inclusions, as identified through the use of EDS mapping of Mn and S, has been demonstrated to serve as a stress concentrator. These inclusions have been observed to initiate micro-pits, which subsequently evolve into visible cracks. Strain contouring analysis reveals that plastic deformation localizes at the crack front (see Figure 14d), thereby accelerating film rupture and chloride ingress. Additionally, kernel average misorientation (KAM) mapping reveals elevated subgrain rotation densities in the immediate vicinity of the crack (see Figure 14e), suggesting the accumulation of geometrically necessary dislocations (GNDs). These GNDs not only accommodate strain but also promote crack tip branching under cyclic film breakdown.

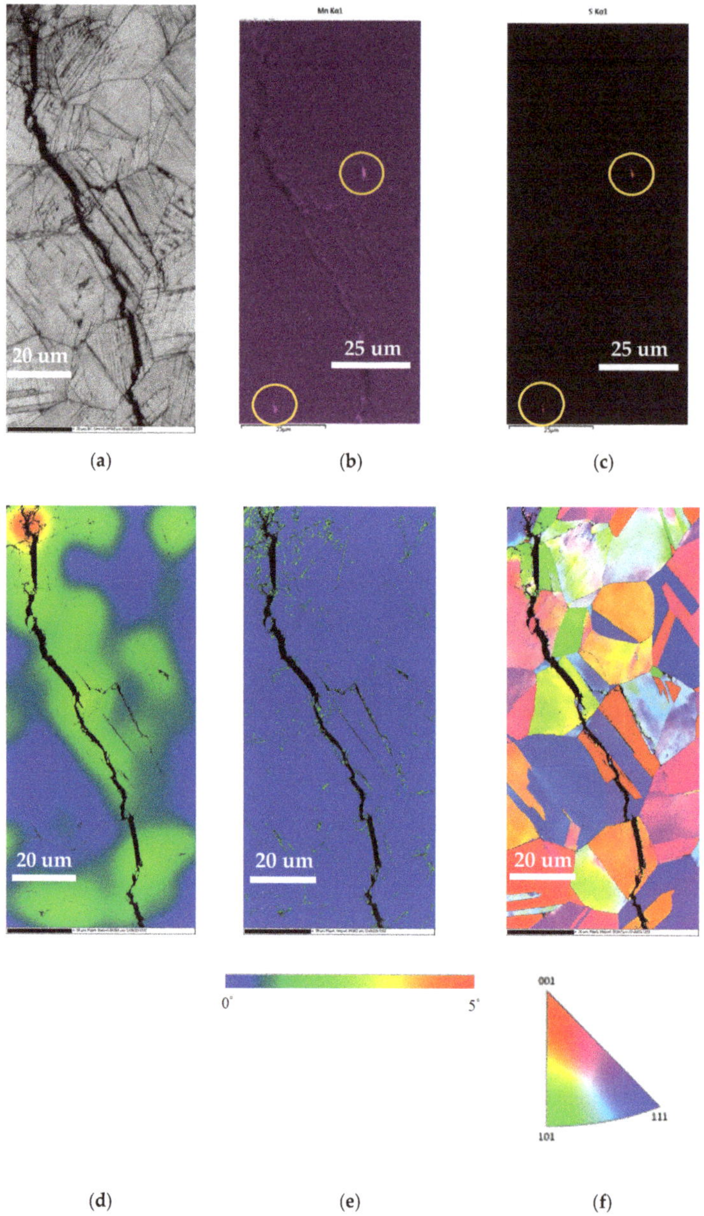

Figure 14. EBSD maps and EDS mapping for the crack areas of the specimen enveloped within the crevice former and deposited at a 1 g/m^2 chloride concentration exposed to 35 °C and 70% RH: (**a**) IQ figure, (**b**) Mn mapping, (**c**) S mapping, (**d**) strain countering map, (**e**) KAM map, and (**f**) inverse pole figure (IPF) map.

4. Discussions

Following an 8000 h test, SCC cracks were observed exclusively on the specimen subjected to white emery spray and tested at a temperature of 45 °C and a 0.1 g/m^2 chloride concentration. Increasing the chloride concentration to 1 g/m^2 resulted in a higher level of corrosion and a greater severity of SCC for the specimens enveloped within the crevice former (see Figures 7 and 12). This was observed in comparison to the specimens subjected to white emery spray (see Figures 6 and 11) at both temperature levels. Factors influencing crevice corrosion initiation include the existence of aggressive chlorine ions, the ambient temperature, and the crevice geometry. Generally, the severity of environmental conditions is directly proportional to the tightness of the crevice and therefore to the probability of corrosion initiation [42]. The presence of deposits, such as dust, on the metal surface has been shown to act as a crevice former. The testing of the PTFE crevice former was conducted by maintaining the specimen and the crevice former in a fixed position through the application of a torque of 1.13 Nm. Conversely, the testing of the white emery deposit was conducted with no external force applied to the sample and the white emery for fixation. Consequently, the crevice was tighter when the specimen was enveloped within the crevice former than when it was sprayed with white emery. Nevertheless, an increase in surface roughness indicates the presence of deeper grooves on the material surface, where a higher concentration of chloride ions occurs, resulting in early corrosion initiation. The roughness of the test specimen was measured at 0.5 μm in the testing of the PTFE crevice former, while the as-machined specimen exhibited a surface roughness of 4.9 μm in the testing of the white emery deposit.

In summary, the 8000 h test demonstrated that an enhancement in the effect of surface roughness was observed when the chloride deposition density for the samples was 0.1 g/m^2. Consequently, the presence of SCC cracks was exclusively observed in specimens that were subjected to white emery spray and tested at a temperature of 45 °C. However, an increase in the chloride deposition density to 1 g/m^2 resulted in crevice tightness becoming the predominant factor, surpassing surface roughness. This observation indicates that specimens enveloped within the crevice former revealed higher levels of corrosion and severity of surface cracks compared to those subjected to white emery spray at both temperature levels. The hypothesis suggested that crevice corrosion is affected by two interrelated factors: surface roughness and crevice tightness. It can be concluded that the SCC initiation of 304L stainless steel was the result of the combined effects of surface roughness and crevice tightness.

5. Conclusions

This study analyzes the characteristics of crevice corrosion behavior of 304L SS subjected to combined ambient temperatures of 35 °C and 45 °C, relative humidities of 45%, 55%, and 70%, chloride concentrations of 0.1 g/m^2 and 1 g/m^2, and test durations of 8000 h and 23,000 h. The specimens were tested with PTFE crevice formers wrapped on the grounded surfaces and white emery deposited on the surface of the as-machined material. The conclusions of the present study are summarized below:

(1) In the 8000 h tests, discontinuous SCC cracks were observed only in the specimens subjected to white emery spray at 45 °C and 70% relative humidity with a 0.1 g/m^2 chloride deposition density; at a 1 g/m^2 chloride deposition density, the PTFE crevice former test specimens developed continuous SCC cracks under identical conditions. The extension of the test to 23,000 h at a 0.1 g/m^2 chloride deposition density resulted in the formation of continuous cracks, even under the presence of white emery deposits. This observation indicates that the SCC initiation threshold is located within the range of 8000 h to 23,000 h.

(2) When environmental conditions are constant, the duration of the tests is observed to affect the morphology of SCC cracks on the specimens. The cracks transition from a discontinuous to a continuous mode as the duration of the test increases. This finding suggests that the morphologically observed evolution from discrete to interconnected cracks is driven by a synergistic effect of increasing local stress intensity under crevice

confinement, progressive chloride ion enrichment leading to passive film breakdown, and microstructural degradation marked by dislocation accumulation.

(3) The specimens with a chloride deposition density of 1 g/m^2 experienced significant crevice corrosion exclusively in regions where white emery was deposited.

(4) The 304L stainless steel specimens exhibited cracking under TGSCC mode, which was confirmed by the EBSD analysis results.

Author Contributions: Conceptualization, C.-P.Y. and K.-C.T.; formal analysis, C.-P.Y.; investigation, C.-P.Y.; writing—original draft, C.-P.Y.; writing—review and editing, C.-P.Y. and J.-Y.H. All authors have read and agreed to the published version of the manuscript.

Funding: This research was supported by the National Science and Technology Council (NSTC) Program (PG11406-0069) and National Science and Technology Program (114-2001-02-28-08) of Taiwan.

Institutional Review Board Statement: Not applicable.

Informed Consent Statement: Not applicable.

Data Availability Statement: The original contributions presented in this study are included in the article. Further inquiries can be directed to the corresponding author.

Conflicts of Interest: The authors declare no conflicts of interest.

References

1. Jeon, C.; Kong, B.S.; Chen, J.; Xiao, Q.; Jang, C. The effect of microstructure of an advanced duplex stainless steel on the pitting corrosion and chloride-induced stress corrosion cracking resistance. *Corros. Sci.* **2024**, *240*, 112489. [CrossRef]
2. Kim, S.; Ahn, K.; Kim, G.; Song, S.W. Synchrotron X-ray fluorescence imaging study on chloride-induced stress corrosion cracking behavior of austenitic stainless steel welds via selective corrosion of δ-ferrite. *Corros. Sci.* **2023**, *218*, 111176. [CrossRef]
3. Spencer, D.T.; Edwards, M.R.; Wenman, M.R.; Tsitsios, C.; Scatigno, G.G.; Chard-Tuckey, P.R. The initiation and propagation pf chloride-induced transgranular stress corrosion cracking (TGSCC) of 304L austenitic stainless steel under atmospheric conditions. *Corros. Sci.* **2014**, *88*, 76–88. [CrossRef]
4. Jeon, C.; Shin, J.H.; Kong, B.S.; Chen, J.; Xiao, Q.; Jang, C.; Kim, Y.J. Development of a duplex stainless steel for dry storage canister with improved chloride-induced stress corrosion cracking resistance. *Nucl. Eng. Technol.* **2024**, *56*, 2131–2140. [CrossRef]
5. Kim, S.; Kim, G.; Kim, C.K.; Song, S.W. Effects of environmental parameters on chloride-induced stress corrosion cracking behavior of austenitic stainless steel welds for dry storage canister application. *Nucl. Eng. Technol.* **2024**, *56*, 317–327. [CrossRef]
6. Qu, H.J.; Tatman, J.; Wharry, J.P. Chloride-induced stress corrosion cracking in Austenitic steels for SNF storage canisters -Recent understanding and advances in mitigation. *J. Nucl. Mater.* **2024**, *596*, 155080. [CrossRef]
7. Zhao, X.Y.; Cheng, C.Q.; Zhang, D.J.; Zhao, Y.N.; Cao, T.S.; Zhong, S.; Zhang, L.; Zhao, J. Effect of U-bending deformation on pitting corrosion of 2205 duplex stainless steel under wet-dry cycling of chloride salt droplets. *Corros. Sci.* **2023**, *218*, 111185. [CrossRef]
8. Dhaiveegan, P.; Elangovan, N.; Nishimura, T.; Rajendran, N. Corrosion behavior of 316L and 304 stainless steels exposed to industrial-marine-urban environment: Field study. *RSC Adv.* **2016**, *6*, 47314–47324. [CrossRef]
9. Burkert, A.; Müller, T.; Lehmann, J.; Mietz, J. Long-term corrosion behavior of stainless steels in marine atmosphere. *Mater. Corros.* **2018**, *69*, 20–28. [CrossRef]
10. Lv, W.; Pan, C.; Su, W.; Wang, Z.; Liu, S.; Wang, C. A study on atmospheric corrosion of 304 stainless steel in a simulated marine atmosphere. *J. Mater. Eng. Perform.* **2015**, *24*, 2597–2604. [CrossRef]
11. Wang, W.Y.; Tseng, Y.S.; Yeh, T.K. Tolerance assessment and crack growth of chloride-induced stress corrosion cracking for Chinshan dry storage system. *Prog. Nucl. Energy* **2022**, *147*, 104210. [CrossRef]
12. Jensen, P.J.; Suffield, S.; Grant, C.L.; Spitz, C.; Hanson, B.; Ross, S.; Durbin, S.; Bryan, C.; Saltzstein, S. Preliminary modeling of chloride deposition on spent nuclear fuel canisters in dry storage relevant to stress corrosion cracking. *Nucl. Technol.* **2022**, *208*, 586–601. [CrossRef]
13. Li, Y.Z.; Wang, X.; Zhang, G.A. Corrosion behavior of 13Cr stainless steel under stress and crevice in 3.5 wt.% NaCl solution. *Corros. Sci.* **2020**, *163*, 108290. [CrossRef]
14. Han, D.; Jiang, Y.; Shi, C.; Deng, B.; Li, J. Effect of temperature, chloride ion and pH on the crevice corrosion behavior of SAF 2205 duplex stainless steel in chloride solutions. *J. Mater. Sci.* **2012**, *47*, 1018–1025. [CrossRef]

15. Meyer, R.M.; Pardini, A.; Cuta, J.; Adkins, H.; Casella, A.; Qiao, A.; Larche, M.R.; Diaz, A.; Doctor, S.R. *NDE to Manage Atmospheric SCC in Canister for Dry Storage of Spent Fuel: An Assessment*; PNNL-22495 401001060; Pacific Northwest National Laboratory: Richland, WA, USA, 2013.
16. Machuca, L.L.; Bailey, S.I.; Gubner, R.; Watkin, E.L.; Ginige, M.P.; Kaksonen, A.H.; Heidersbach, K. Effect of oxygen and biofilms on crevice corrosion of UNS S31803 and UNS N08825 in natural seawater. *Corros. Sci.* **2013**, *67*, 242–255. [CrossRef]
17. Kennel, G.F.; Evitts, R.W.; Heppner, K.L. A Critical crevice solution and IR drop crevice corrosion model. *Corros. Sci.* **2008**, *50*, 1716–1725. [CrossRef]
18. Schoell, R.; Xi, L.; Zhao, Y.; Wu, X.; Yu, Z.; Kenesei, P.; Almer, J.; Shayer, Z.; Kaoumi, D. In situ synchrotron X-ray tomography of 304 stainless steels undergoing chloride-induced stress corrosion cracking. *Corros. Sci.* **2020**, *170*, 108687. [CrossRef]
19. Yeom, H.; Dabney, T.; Pocquette, N.; Ross, K.; Pfefferkorn, F.E.; Sridharan, K. Cold spray deposition of 304L stainless steel to mitigate chloride-induced stress corrosion cracking in canisters for used nuclear fuel storage. *J. Nucl. Mater.* **2020**, *538*, 152254. [CrossRef]
20. Wu, X. On residual stress analysis and microstructural evolution for stainless steel type 304 spent nuclear fuel canisters weld joint: Numerical and experimental studies. *J. Nucl. Mater.* **2020**, *534*, 152131. [CrossRef]
21. Mayuzumi, M.; Arai, T.; Hide, K. Chloride induced stress corrosion cracking of type 304 and 304L stainless steels in air. *Zair. Kankyo* **2003**, *52*, 166–170. [CrossRef]
22. Tani, J.I.; Mayuzumi, M.; Hara, N. Stress corrosion cracking of stainless-steel canister for concrete cask storage of spent fuel. *J. Nucl. Mater.* **2008**, *379*, 42–47. [CrossRef]
23. Guo, L.; Mi, N.; Mohammed-Ali, H.; Ghahari, M.; Plessis, A.D.; Cook, A.; Street, S.; Reinhard, C.; Atwood, R.C.; Rayment, T.; et al. Effect of mixed salts on atmospheric corrosion of 304 stainless steel. *J. Electrochem. Soc.* **2019**, *166*, C3010–C3014. [CrossRef]
24. Dong, P.; Scatigno, G.G.; Wenman, M.R. Effect of salt composition and microstructure on stress corrosion cracking of 316L austenitic stainless steel for dry storage canisters. *J. Nucl. Mater.* **2021**, *545*, 152572. [CrossRef]
25. Scatigno, G.G.; Dong, P.; Ryan, M.P.; Wenman, M.R. The effects of salt loading on chloride-induced stress corrosion cracking of 304L austenitic stainless steel under atmospheric conditions. *Materialia* **2019**, *8*, 100509. [CrossRef]
26. Shoji, S.; Ohnaka, N.; Furutani, Y.; Saitoh, T. Effects of relative humidity on atmospheric stress corrosion cracking of stainless steels. *Corros. Eng.* **1986**, *35*, 559–565. [CrossRef]
27. Scatigno, G.G.; Ryan, M.P.; Giuliani, F.; Wenman, M.R. The effects of prior cold work on the chloride stress corrosion cracking of 304 L austenitic stainless steel under atmospheric condition. *Mater. Sci. Eng. A* **2016**, *668*, 20–29. [CrossRef]
28. Cook, A.B.; Lyon, S.B.; Stevens, N.P.C.; Gunther, M.; McFiggans, G.; Newman, R.C.; Engelberg, D.L. Assessing the risk of under-deposit chloride-induced stress corrosion cracking in austenitic stainless steel nuclear waste containers. *Corros. Eng. Sci. Technol.* **2014**, *49*, 529–534. [CrossRef]
29. Prosek, T.; Iverson, A.; Taxsen, C.; Thierry, D. Low-temperature stress corrosion cracking of stainless steels in the atmosphere in the presence of chloride deposits. *Corrosion* **2009**, *65*, 105–117. [CrossRef]
30. Ornek, C.; Engelberg, D.L. Toward understanding the effects of strain and chloride deposition density on atmospheric chloride-induced stress corrosioncracking of type 304 austenitic stainless steel under $MgCl_2$ and $FeCl_3$: $MgCl_2$ Droplets. *Corrosion* **2009**, *75*, 167–182. [CrossRef]
31. Oberson, G.; Dunn, D.; Mintz, T.; He, X.; Pabalan, R.; Miller, L. US NRC-sponsored research on stress corrosion cracking susceptibility of dry storage canister materials in marine environments-13344. In Proceedings of the WM2013 Conference, Phoenix, AZ, USA, 24–28 February 2013.
32. He, X.; Mintz, T.S.; Pabalan, R.; Miller, L.; Oberson, G. *Assessment of Stress Corrosion Cracking Susceptibility for Austenitic Stainless Steels Exposed to Atmospheric Chloride and Non-Chloride Salts*; NUREG/CR-7170; U.S. Nuclear Regulatory Commission: Rockville, MD, USA, 2014.
33. Shirai, K.; Tani, J.; Saegusa, T. *Study on Interim Storage of Spent Nuclear Fuel by Concrete Cask for Practical Use—Feasibility Study on Prevention of Chloride Induced Stress Corrosion Cracking for Type304L Stainless Steel Canister*; CRIEPI N10035; Central Research Institute of Electric Power Industry: Chiyoda-Ku, Japan, 2014.
34. Takeda, H.; Wataru, M.; Shirai, K.; Saegusa, T. Heat removal verification tests using concrete casks under normal condition. *Nucl. Eng. Des.* **2008**, *238*, 1196–1205. [CrossRef]
35. Qiao, L.J.; Gao, K.W.; Volinsky, A.A.; Li, X.Y. Discontinuous surface cracks during stress corrosion cracking of stainless steel single crystal. *Corros. Sci.* **2011**, *53*, 3509–3514. [CrossRef]
36. Masuda, H. SKFM observation of SCC on SUS304 stainless steel. *Corros. Sci.* **2007**, *49*, 120–129. [CrossRef]
37. Acharyya, S.G.; Khandelwal, A.; Kain, V.; Kumar, A.; Samajdar, I. Surface working of 304L stainless steel: Impact on microstructure, electrochemical behavior and SCC resistance. *Mater. Charact.* **2012**, *72*, 68–76. [CrossRef]
38. *D1141-98*; Standard Practice for the Preparation of Substitute Ocean Water. ASTM International: West Conshohocken, PA, USA, 2013.
39. Yeh, C.P.; Tsai, K.C.; Huang, J.Y. The effect of deposited dust on SCC and crevice corrosion of AISI 304L stainless steel in saline environment. *Materials* **2021**, *14*, 6834. [CrossRef]

40. Truman, J.E. The influence of chloride content, pH and temperature of test solution on the occurrence of stress corrosion cracking with austenitic stainless steel. *Corros. Sci.* **1977**, *17*, 737–746. [CrossRef]
41. Li, J.X.; Chu, W.Y.; Wang, Y.B.; Qiao, L.J. In situ TEM study of stress corrosion cracking of austenitic stainless steel. *Corros. Sci.* **2003**, *45*, 1355–1365. [CrossRef]
42. Oldfield, J.W. Test techniques for pitting and crevice corrosion resistance of stainless steels and nickel-base alloys in chloride-containing environments. *Int. Mater. Rev.* **1987**, *32*, 153–172. [CrossRef]

Disclaimer/Publisher's Note: The statements, opinions and data contained in all publications are solely those of the individual author(s) and contributor(s) and not of MDPI and/or the editor(s). MDPI and/or the editor(s) disclaim responsibility for any injury to people or property resulting from any ideas, methods, instructions or products referred to in the content.

materials MDPI

Article

Titanium Modulated the Occurrence States and Strain Aging Resistance of Residual Element Nitrogen in Scrap-Based Low-Alloy Steels

Yuhe Huang [1,2,*,†], Haisheng Yang [1,†], Jun Lu [1,2], Jing Wang [3], Bicao Peng [3], Junheng Gao [1,2], Haitao Zhao [1,2], Honghui Wu [1,2], Chaolei Zhang [1,2], Shuize Wang [1,2,*] and Xinping Mao [1,2]

1 Institute for Carbon Neutrality, University of Science and Technology Beijing, Beijing 100083, China
2 Institute for Steel Sustainable Technology, Liaoning Academy of Materials, Shenyang 110004, China
3 Hunan Iron and Steel Group Co., Ltd., Changsha 410100, China
* Correspondence: huangyuhe@ustb.edu.cn (Y.H.); wangshuize@ustb.edu.cn (S.W.)
† These authors contributed equally to this work.

Abstract

The steel industry is responsible for 7–9% of global CO_2 emissions. Shifting from primary iron ore to recycled scrap in electric arc furnace (EAF) steelmaking offers significant decarbonization potential, reducing carbon intensity by 60–70%. However, increased scrap use in EAF operations leads to higher nitrogen absorption, which can degrade mechanical properties. Nitrogen dissolves into molten steel, where it forms Cottrell atmospheres at dislocations in the following processing steps, intensifying strain aging and reducing ductility. This study establishes a precipitation criterion based on the TiN solubility product to prevent harmful liquid TiN formation, enabling effective nitrogen fixation via fine TiN precipitates (5–20 nm). Multiscale characterization techniques, such as TEM and EBSD, show that Ti reduces the number of mobile N atoms by 60–70%, evidenced by a 50–65% decrease in Snoek/SKK peak intensities. Excessive titanium can refine ferrite grain size and prevents harmful TiN inclusions. Titanium microalloying presents a cost-effective, sustainable strategy to reduce strain aging in scrap-rich EAF steels, enabling more sustainable steel production without sacrificing material properties.

Keywords: residual element; nitrogen; strain aging sensitivity; low-alloy steel; titanium microalloying; scrap-based steel

1. Introduction

Iron and steel manufacturing generates nearly one-tenth of global anthropogenic CO_2 emissions, driving imperative decarbonization efforts. A cornerstone strategy involves substituting virgin iron ore with recycled ferrous scrap in metallurgical processes—each tonne of scrap utilization avoids approximately 1.5 tonnes of CO_2 equivalent emissions [1]. This circular approach is optimally implemented through electric arc furnace (EAF) technology, where scrap constitutes more than 80% of charge materials. The EAF route demonstrates fundamentally superior carbon efficiency, exhibiting 60–70% lower greenhouse gas intensity relative to coal-intensive conventional blast furnace-converter (BF-BOF) pathways [2]. However, inherent metallurgical constraints impede broader EAF adoption, with nitrogen management constituting a primary technological challenge. In EAF operations, the intense electric arcs and high temperatures promote the dissociation of atmospheric N_2 into highly reactive atomic nitrogen (N), which readily dissolves into molten steel [3].

Consequently, EAF steels often exhibit elevated levels of residual N, posing a major barrier to the production of high-quality, low-N steels.

Nitrogen's influence on steel performance manifests as a state-dependent duality governed by concentration and microstructural interactions. At controlled levels, interstitial N enhances mechanical properties through solid-solution strengthening and precipitation hardening mechanisms [4]. Fine-scale nitride formation restricts grain boundary mobility during thermomechanical processing, refining microstructure concurrently. This principle is exploited in austenitic stainless systems where nitrogen alloying (0.1–0.25 wt%) stabilizes γ-phase crystallography, simultaneously enhancing pitting resistance and yield strength [5,6]. Conversely, excessive nitrogen dissolution—endemic to EAF processing with concentrations typically reaching 60–100 ppm—initiates detrimental phenomena including dislocation locking and intergranular embrittlement. These mechanisms fundamentally degrade mechanical integrity through strain aging susceptibility and cleavage fracture initiation.

These detrimental effects originate from nitrogen's interstitial mobility and segregation kinetics. Solute N atoms preferentially diffuse to crystalline defects during cooling or deformation, forming Cottrell atmospheres that pin dislocations and creating brittle nitridic intergranular films (e.g., ε-$Fe_{2-3}N$). While such segregation elevates yield strength through dislocation locking, it catastrophically compromises ductility and Charpy impact energy [7]. Furthermore, nitrogen supersaturation elevates the ductile–brittle transition temperature (DBTT) through dislocation immobilization mechanisms, exacerbating cold brittleness in service environments [8]. Additional failure modes include subsurface porosity from nitrogen repartitioning during solidification, macrosegregation-induced yield strength reduction, stress-concentrating TiN/AlN inclusions nucleating microcracks, and accelerated fatigue crack initiation at nitride–matrix interfaces [9,10]. The thermodynamic constraints of EAF processing render precise N management both technically challenging and economically prohibitive, frequently resulting in solute concentrations surpassing critical thresholds (typically > 60 ppm) [11,12]. Conventional mitigation approaches including vacuum processing and reactive gettering agents, incurring substantial operational costs that limit scalability for commodity steel production [11–13]. This challenge acutely impacts low-alloy structural grades exemplified by SAPH440, where the essential strength-toughness-weldability triad meets critical demand in automotive chassis components and structural frameworks. The scrap-intensive EAF route inherently elevates residual nitrogen to 60–100 ppm, intensifying dynamic strain aging phenomena during subsequent thermomechanical forming operations such as coil leveling. To reconcile decarbonization imperatives with stringent performance specifications, novel cost-effective approaches for modulating nitrogen's occurrence state through microstructural engineering become indispensable.

Microalloying with strong nitride-forming elements presents a viable pathway for N management, where additions of niobium (Nb), vanadium (V), or titanium (Ti) immobilize interstitial nitrogen through precipitation of thermodynamically stable MN-type compounds. For instance, Nb and V are known to form fine carbonitrides that slow recrystallization and refine ferrite grains [14–16]. Titanium, due to its high affinity for nitrogen and lower cost compared to Nb and V, offers a promising alternative for nitride stabilization and strain aging suppression in scrap-based EAF steels [17–19].

Despite this potential, significant scientific uncertainties hinder optimal implementation. Critical among these are the undefined processing parameters required to avoid deleterious liquid TiN formation during solidification, unresolved relationships between Ti/N stoichiometry and nitrogen partitioning behavior, and competing microstructural consequences between strain aging suppression and precipitation evolution. This study systematically addresses these knowledge gaps by investigating Ti additions in scrap-based SAPH440 steel through integrated computational and experimental methodologies. Using integrated thermodynamic simulation, multiscale characterization, and mechanical testing, precipitation criteria were established to prevent harmful inclusions, quantify nitrogen's transformation across solute/cluster/precipitate states, and elucidate titanium's mechanistic role in enhancing strain aging resistance while optimizing strength-ductility synergy. The scientific novelty of this work lies in the establishment of a TiN precipitation criterion for scrap-based EAF steels, coupled with multiscale evidence of Ti-mediated nitrogen immobilization and strain aging suppression, providing a cost-effective pathway for high-performance, low-carbon steel production.

2. Materials and Methods

The experimental material in this study was commercial-grade SAPH440 low-alloy steel produced via the scrap-based electric arc furnace (EAF) process. In conventional converter steelmaking, the N content in the final molten steel typically stabilizes at approximately 40 ppm. To systematically investigate the role of Ti in regulating N behavior, a series of samples with controlled Ti/N ratios were designed. Although Ti effectively fixes N by forming stable compounds, its high affinity for N, sulfur (S), and oxygen (O) may inadvertently promote inclusion formation. Notably, TiN precipitates formed in the liquid phase are particularly detrimental due to their coarse particle size (>1 μm) and sharp cubic morphology, which act as stress concentrators and markedly degrade material ductility. To suppress the formation of these harmful precipitates, the solubility product of liquid TiN was adopted as the key design criterion for alloy composition optimization [14–16]:

$$\lg([\%Ti][\%N])_L = 5.9 - 16586/T \quad (1)$$

$$\lg([\%Ti][\%N])_\delta = 5.56 - 17205/T \quad (2)$$

Using Equations (1) and (2), where [%Ti] and [%N] represent the mass percentages of titanium and nitrogen in the steel, respectively, and T is the absolute temperature in Kelvin (K), the critical concentrations of Ti and N required to prevent the formation of liquid TiN were determined (Figure 1). The solid curves represent the calculated solubility products of TiN at different temperatures (1450 °C, 1480 °C, and 1520 °C). The 1520 °C curve corresponds to the liquid state condition. As shown in Figure 1, for N levels below 100 ppm, Ti additions of $\leq$0.03 wt.% effectively prevent liquid TiN formation, allowing the N to precipitate in the solid phase. Based on this criterion, four experimental steels with controlled Ti/N ratios were designed: 80N (reference, with 80 ppm N), 0.03Ti-60N (with 0.03 wt.% Ti and 60 ppm N), 0.015Ti-60N (with 0.015 wt.% Ti and 60 ppm N), and 0.015Ti-80N (with 0.015 wt.% Ti and 80 ppm N). Additionally, to ensure the steel's strength remains stable after Ti addition, the Mn content in the Ti microalloyed samples was reduced from 1.4 to 1 wt.%. The chemical compositions of these experimental steels are presented in Table 1.

Table 1. Experimental steel measured chemical composition (wt.%).

Sample	C	Si	Mn	Ti	N	Fe
80N	0.052	0.19	1.42	0	0.0079	Bal.
0.03Ti-60N	0.055	0.19	1.04	0.026	0.0065	Bal.
0.015Ti-80N	0.049	0.19	1.06	0.012	0.0081	Bal.
0.015Ti-60N	0.052	0.19	1.06	0.016	0.0062	Bal.

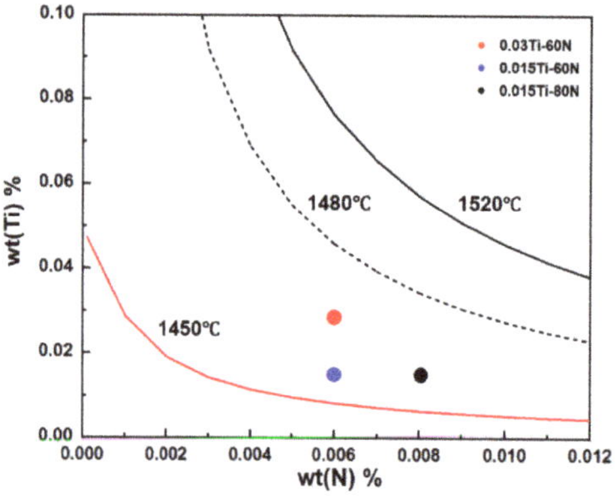

Figure 1. The solubility product curve of TiN in liquid and solid steel.

The investigated material was subjected to industrially relevant thermo-mechanical processing through laboratory hot-rolling simulation. Initial homogenization conducted at 1200 °C for 60 min to ensure chemical uniformity. Subsequent hot-rolling comprised five consecutive deformation passes in the temperature range of 1050 °C to 830 °C, with a thickness reduction from 40 mm to 3 mm (approx. 25% reduction per pass). Finish rolling was conducted at 830 °C, followed by accelerated cooling to 550 °C and simulated coiling. The chemical composition of the experimental material was determined by wet chemical analysis. The measured composition is presented in Table 1. The critical phase transformation temperatures were determined using thermodynamic calculations performed with the Thermo-Calc software and the TCFE9 database. For microstructural characterization, longitudinal sections were mechanically prepared through sequential grinding and diamond polishing, with final microstructural revelation using 2% nital etching. Comprehensive analysis employed a scanning electron microscopy (TESCAN Mira LMS, TESCAN, Brno, Czech Republic) equipped with a secondary electron (SE) detector, coupled with electron backscatter diffraction (Oxford Symmetry S2, Oxford Instruments, Oxford, UK) at a 0.2 μm step resolution, utilizing specimens prepared by electrochemical polishing in 5% perchloric acid/ethanol solution. The grain sizes were measured from SEM images using the linear intercept method according to the ASTM E112 standard. Complementary transmission electron microscopy (JEOL JEM 200F, JEOL Ltd., Tokyo, Japan) analysis was conducted on electron-transparent foils prepared through mechanical thinning to ~50 μm followed by twin-jet electropolishing. Mechanical evaluation was performed via uniaxial tensile testing at ambient temperature. To ensure statistical reliability, tests were conducted with three repetitions per condition. The testing protocol involved applying pre-strains of 2% and 10%, followed by artificial aging at 250 °C for 60 min to simulate long-term service conditions. Fracture resistance was assessed via Charpy V-notch impact testing (Zwick/Roell ZBC2452-B, ZwickRoell GmbH & Co. KG, Ulm, Germany). The N occurrence state was studied using an internal friction apparatus (MFP-1000, Institute of Metal Research, Chinese Academy of Sciences, Shenyang, China).

3. Results and Discussions

3.1. Effect of Ti Addition on Microstructure and Mechanical Properties in High Residual N Steel

The microstructures of the experimental steels with varying Ti/N ratios were analyzed using scanning electron microscopy (SEM). Figure 2 presents the SEM images of the experimental steels. All the steels exhibited a microstructure consisting primarily of ferrite. The ferrite grains were found to be equiaxed and irregular in shape, with boundaries that were often curved and non-linear. Microstructural analysis confirmed that the 80N steel primarily consists of ferrite and small amount of pearlite, with no significant evidence of bainitic transformation under the given processing conditions (Figure 2c). In contrast, the Ti microalloyed steels mainly consisted of blocky ferrite grains, with only a small amount of pearlite (denoted as 'P' in Figure 2a). This change can be attributed to the addition of Ti, which promotes the formation of carbides and nitrides during solid-state precipitation upon cooling. During this process, large amounts of C and N are consumed from the solid solution. As C is a critical component in the formation of cementite (Fe_3C), the reduction in C content significantly decreases cementite formation, thereby reducing the formation of pearlite.

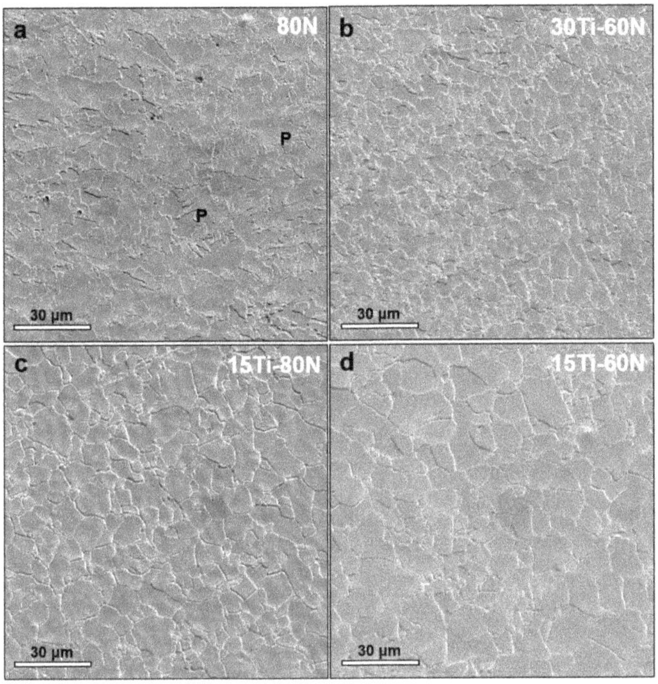

Figure 2. SEM images of experimental steels with varying N and Ti contents: (**a**) 80N, (**b**) 0.03Ti-60N, (**c**) 0.015Ti-80N, and (**d**) 0.015Ti-60N.

Although the Ti/N ratio changes, the morphology of the ferrite remains relatively constant, while changes in grain size are more noticeable. The grain sizes of the 80N, 0.03Ti-60N, 0.015Ti-80N, and 0.015Ti-60N experimental steels were found to be 4.51 $\pm$ 0.23 µm, 4.83 $\pm$ 0.31 µm, 5.97 $\pm$ 0.35 µm, and 6.31 $\pm$ 0.28 µm, respectively. The 0.03Ti-60N composition, exceeding the stoichiometric Ti/N ratio of 3.42 [17], enabled secondary TiC precipitation. These nanoscale carbides exerted potent boundary pinning effects through Zener drag mechanisms, explaining the enhanced grain refinement beyond nitrogen's solid solution effects. Conversely, sub-stoichiometric 0.015Ti variants lacked sufficient titanium for carbide formation, resulting in coarser microstructures with diminished precipitate-mediated grain boundary restraint.

Figure 3 displays both the Kernel Average Misorientation (KAM) and grain boundary maps of the experimental steels. The KAM values for the 80N, 0.03Ti-60N, 0.015Ti-80N, and 0.015Ti-60N steels were 0.33°, 0.19°, 0.14°, and 0.14°, respectively. While the KAM values did not show a strong monotonic trend, the sample with the highest Ti content (0.03Ti-60N) exhibited a higher KAM value compared to the samples with lower Ti content (0.015Ti-80N and 0.015Ti-60N), suggesting that a higher density of fine precipitates (e.g., TiC in 0.03Ti-60N) may impede dislocation recovery, even as the overall dislocation density is reduced compared to the Ti-free steel. Compared to the Ti-free 80N steel, the Ti-containing steels exhibited lower KAM values, which can be attributed to the relatively lower dislocation density in the blocky ferrite microstructure. Regarding the grain boundary characteristics, boundaries were categorized based on their orientation difference. Low-angle grain boundaries (LAGBs) are defined by an orientation difference of 2°–15°, while high-angle grain boundaries (HAGBs) have an orientation difference greater than 15°. From Figure 3, it is evident that there is minimal variation in the fraction of HAGBs between the 0.03Ti-60N, 0.015Ti-80N, and 0.015Ti-60N steels. However, the 80N steel, which lacks Ti, shows the highest fraction of LAGBs. This is likely due to the absence of irregular ferrite or pearlite structures in the Ti-containing steels. The more regular, blocky ferrite morphology in these alloys tends to favor the formation of HAGBs over LAGBs.

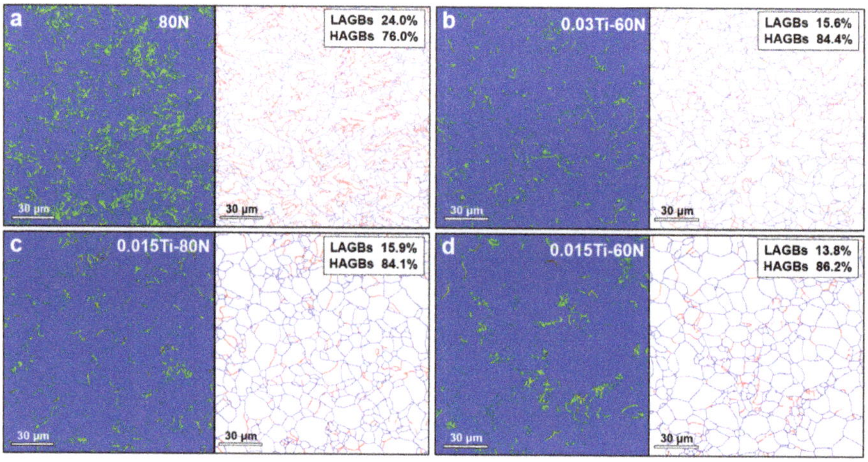

Figure 3. KAM and grain boundary images of experimental steels with varying N and Ti contents: (**a**) 80N, (**b**) 0.03Ti-60N, (**c**) 0.015Ti-80N, and (**d**) 0.015Ti-60N.

To investigate the influence of Ti addition on the phase transformation process and its subsequent effect on the microstructure, the phase transition points of the experimental steels with varying Ti/N ratios were calculated through thermos-calc. The results are presented in Table 2 and Figure 4. Compared to the Ti-free experimental steels, the Ti-containing steels exhibited a decrease in both the austenite start temperature (Ac1) and the ferrite finish temperature (Ar3). Conversely, the austenite finish temperature (Ac3) and ferrite start temperature (Ar1) increased. These findings indicate that the addition of Ti expands the phase transformation region between austenite and ferrite. Furthermore, Ti effectively mitigates the influence of N on the phase transformation temperatures of the experimental steels. By fixing the N content, Ti raises the ferrite transformation temperature and reduces the undercooling temperature during the hot rolling process. These two combined effects lead to a microstructure predominantly consisting of blocky ferrite in the Ti-containing experimental steels. From a phase transformation perspective, this behavior promotes an increase in grain size, as the wider transformation zone encourages the formation of larger grains.

Table 2. Phase transformation temperature of experimental steels.

Sample	Ac1/°C	Ac3/°C	Ar1/°C	Ar3/°C
80N	654	842.5	842.5	654
0.03Ti-60N	532.7	859.6	859.6	532.7
0.015Ti-80N	526.8	858	858	526.8
0.015Ti-60N	521.2	858.8	858.8	521.2

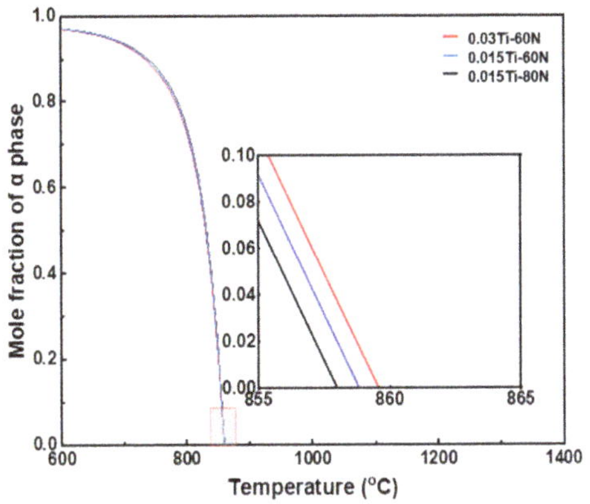

Figure 4. Evolution of the ferrite phase with temperature in experimental steels with varying Ti/N ratio.

Although the 0.03Ti-60N sample exhibits the highest Ar1 (Figure 4), it has the smallest grain size among the Ti-alloyed samples. This can likely be attributed to the pinning effect of the formed particles. Precipitation analysis for steels with different Ti/N ratios is shown in Figure 5a–c. The calculated MN-type precipitates (where M represents microalloying elements such as Ti and Nb, and N represents interstitial elements like C and N) are predominantly TiN for both the 0.015Ti-80N and 0.015Ti-60N samples. Additionally, the calculated liquidus and solidus temperatures for the 0.03Ti-60N, 0.015Ti-80N, and 0.015Ti-60N samples are presented in Table 3. The liquidus temperatures are 1498.6 °C, 1525.2 °C, and 1525.3 °C, respectively, while the corresponding solidus temperatures are 1525.1 °C, 1498.5 °C, and 1490.0 °C. As the total Ti plus N content increases, the liquidus temperature rises, and the solidus temperature decreases, which results in a broader solid–liquid interval. The predicted TiN precipitation temperatures for 0.03Ti-60N, 0.015Ti-80N, and 0.015Ti-60N are 1511.9 °C, 1505.3 °C, and 1504.8 °C, respectively. These results show a positive correlation between the TiN precipitation temperature and the combined Ti and N content.

Table 3. Transformation temperatures of experimental steels with varying Ti/N ratios.

Sample	Precipitation Point (°C)	Liquidus Point (°C)	Solidus Point (°C)
0.015Ti-80N	1505.3	1525.2	1498.5
0.015Ti-60N	1504.8	1525.3	1490
0.03Ti-60N	1511.9	1498.6	1525.1

In Figure 5d–f, representative micrographs of TiN in the three Ti-bearing steels are presented. Coarse, faceted TiN particles were detected in all samples, but their number density was very low, with sizes ranging from 50 to 100 nm. Interestingly, fine TiN precipitates, with sizes between 5 and 20 nm (inset of Figure 5e), were also observed. Given the low levels of Ti and N, and the fact that the computed TiN precipitation temperatures are only slightly above the solidus, it can be inferred that the 50–100 nm TiN particles are primarily liquid-borne precipitates, while the finer 5–20 nm TiN precipitates formed by solid-state precipitation during or after solidification. This interpretation aligns with previous studies, which report that liquid-precipitated TiN tends to be coarser and may impair ductility when present in high quantities [19–23]. In addition to TiN, sub-sized TiC precipitates (~5 nm) were occasionally observed in the 0.03Ti-60N sample. This observation aligns with thermodynamic predictions for 0.03Ti-60N (TiC present) and 0.015Ti-80N (TiC absent) shown in Figure 5a,c. However, for the 0.015Ti-60N sample (Figure 5e), sparse TiC particles were experimentally detected, even though the calculation predicted no TiC formation. The particle density in this alloy was much lower than in the 0.03Ti-60N sample. As predicted by the thermodynamic calculations in Figure 5a–c, the microscopy results broadly support the thermodynamic predictions, while the limited presence of TiC in the 0.015Ti-60N alloy suggests a narrow processing window that allows incipient TiC formation, which goes beyond the equilibrium prediction.

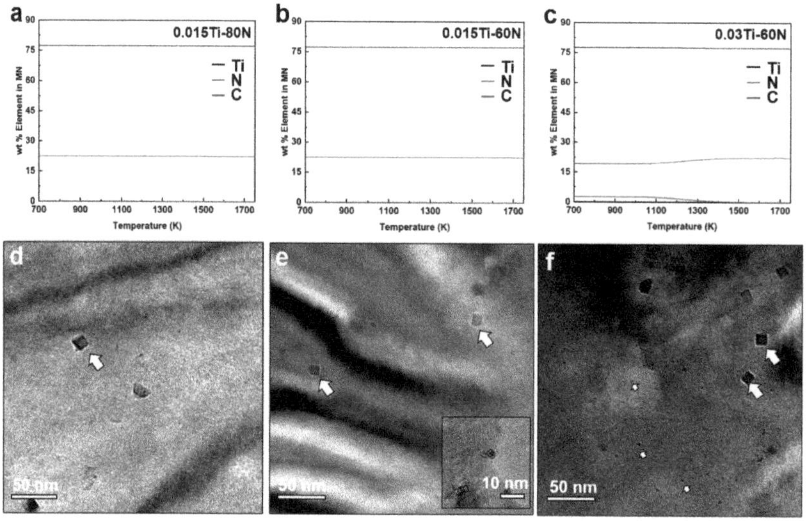

Figure 5. Evolution of the ferrite phase with temperature in experimental steels with different Ti/N contents. (**a**–**c**) Precipitation behavior of steels with varying Ti/N ratios. (**d**–**f**) Representative micrographs showing TiN precipitates in the three Ti-bearing steels.

Internal friction spectroscopy was employed to investigate Ti's influence on N occurrence states across systematically varied Ti/N stoichiometries. Characteristic damping peaks in the experimental spectra correspond to distinct interstitial relaxation mechanisms, the Snoek peak at approximately 340 K originates from stress-induced reorientation of interstitial atoms within body-centered cubic iron lattices, while the Snoek-Kê-Köster (SKK) peak near 520 K arises from interactions between dislocations and interstitial solutes [18]. These relaxation phenomena dissipate vibrational energy (a form of mechanical energy lost as heat during oscillatory stress) through atomistic processes including strain-induced solute ordering, dislocation kink migration, and grain boundary sliding mechanisms. Given the consistent carbon concentrations (0.04–0.06 wt%) across all specimens and analogous octahedral site occupation behaviors of C and N interstitials, variations in Snoek peak amplitude provide direct qualitative assessment of free N content [18]. Concurrently, SKK peak intensity exhibits dual dependence on mobile interstitial concentration and dislocation density, serving as an indicator of Cottrell atmosphere density—complexes where dislocations become pinned by interstitial solute clouds. Figure 6 demonstrates systematic attenuation of both relaxation signatures with Ti additions while their corresponding critical parameters and activation energies are summarized in Tables 4 and 5, respectively. The 0.03Ti-60N steel exhibited a 28% lower Snoek amplitude than the 80N reference, indicating substantial reduction in free N. SKK intensity concurrently decreased by 34%, reflecting diminished dislocation–interstitial interactions. Both peaks progressively shifted toward lower temperatures (15–20 K reduction) with increasing Ti/N ratios, accompanied by 18–22% reductions in calculated activation energies. These phenomena collectively demonstrate Ti's efficacy in immobilizing interstitial species through multiple mechanisms, preferential TiN nucleation consumes free nitrogen, substitutional-interstitial coupling reduces solute mobility, and precipitation-mediated dislocation pinning decreases dislocation density. The attenuated Cottrell atmosphere density evidenced by SKK suppression directly correlates with reduced strain aging susceptibility, which is a critical advancement for scrap-based EAF steels requiring precise interstitial management. Peak shifts further confirm Ti's role in restricting interstitial mobility, with this stabilizing effect intensifying proportionally with Ti content elevation.

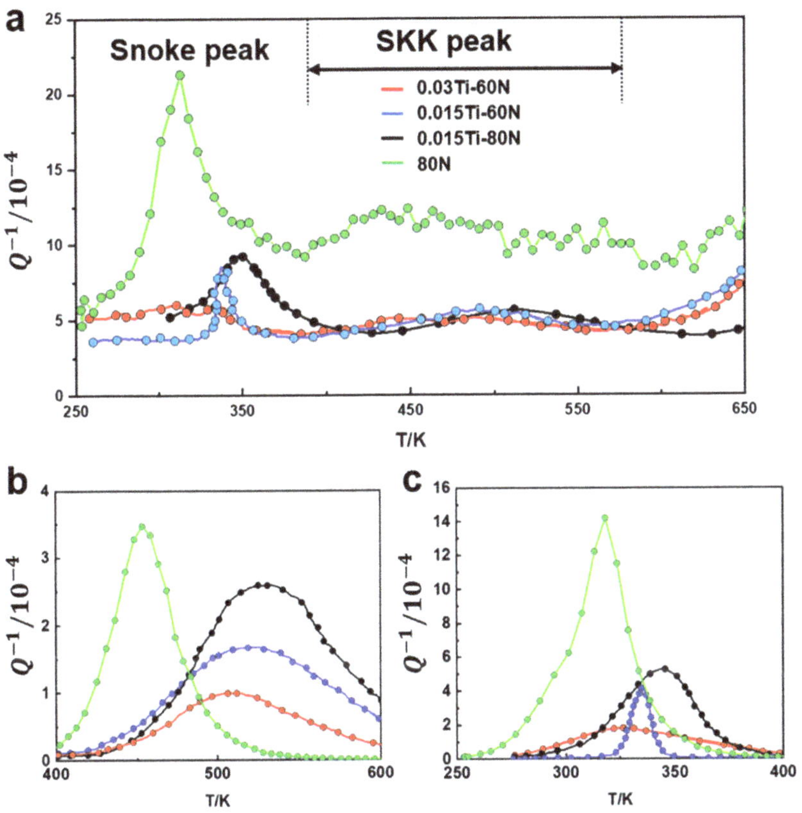

Figure 6. The internal friction experimental results for the experimental steels are presented as follows: (**a**) Original curve, (**b**) Snoek peak, and (**c**) SKK peak.

Table 4. The parameters and activation energy of the Snoek peak in experimental steels with different Ti/N ratios.

Sample	Peak Value	Peak Temperature (K)	Frequency (Hz)	Activation Energy (eV)
80N	0.0142	338	2.388	0.8749
0.015Ti-80N	0.0052	346	2.453	0.8941
0.015Ti-60N	0.0043	336	2.574	0.8661
0.03Ti-60N	0.0018	325	2.552	0.837

Table 5. The parameters and activation energy of the SKK peak in experimental steels with different Ti/N ratios.

Sample	Peak Value	Peak Temperature (K)	Frequency (Hz)	Activation Energy (eV)
80N	0.0035	453	2.431	1.1815
0.015Ti-80N	0.0026	536	2.499	1.4045
0.015Ti-60N	0.0017	522	2.488	1.3668
0.03Ti-60N	0.001	513	2.431	1.3435

3.2. *Effect of Ti Addition on Mechanical Properties and Strain Aging Behavior*

The effect of nitrogen occurrence states on the mechanical properties was evaluated from the tensile stress–strain curves (Figure 7) and the corresponding mechanical properties, including yield strength, ultimate tensile strength, and elongation (Table 6). As the total Ti and N content in the experimental steels increases, there is a general rise in upper yield strength, lower yield strength, ultimate tensile strength, and yield-to-tensile ratio. However, elongation tends to decrease gradually. Additionally, with an increase in the Ti/N ratio, the yield platform length becomes progressively shorter and displays oscillatory behavior. This phenomenon typically indicates the repeated pinning and depinning of dislocations by Cottrell atmospheres. Therefore, the shortened yield platform length is likely related to the reduction in interstitial atoms in the experimental steels. The three experimental steels with different Ti/N ratios mostly meet the required strength and plasticity of SAPH440 low-alloy steel. The 0.03Ti-60N experimental steel has slightly lower plasticity than the required 30%, while the 0.015Ti-60N and 0.015Ti-80N experimental steels exhibit tensile strengths slightly below the required 440 MPa. In contrast, the 80N sample shows higher strength than the standard requirement, which may be attributed to its irregular, elongated grain structure and reduced grain size.

Table 6. Summary of mechanical properties of experimental steels.

Sample	YS/MPa	UTS/MPa	Elongation/%
80N	415 ± 1.7	490 ± 3.2	26.8 ± 0.8
0.03Ti-60N	374 ± 3.5	447 ± 4.9	28.8 ± 1.2
0.015Ti-80N	376 ± 2.8	438 ± 8	30.1 ± 0.9
0.015Ti-60N	364 ± 4.2	433 ± 6.1	31.9 ± 1.4

To simulate industrial processing conditions, including uncoiling (1–2% strain), leveling, and potential long-term storage, as well as end-user cold forming operations (5–10% strain), strain aging behavior was evaluated at 2% and 10% pre-strain levels. Artificial aging at 250 °C for 1 h accelerated natural aging equivalence to approximately one year of ambient exposure. Figure 8 presents tensile responses of 80N and Ti-microalloyed variants after pre-strain and aging treatments. Unprestrained steels exhibited negligible property changes post-aging except for the 80N reference. Following 2% pre-strain and aging, the 80N steel underwent profound mechanical transformation, in which the serrated yield plateau was replaced by a distinct two-stage yielding profile (indicated by yellow arrow, Figure 8a), signaling modified dislocation-solute interactions. This manifested as a dramatic yield strength elevation from 390 MPa to 537 MPa (+38%), coupled with critical ductility loss (15.8% vs. SAPH440's $\geq$ 30% requirement). The strain aging sensitivity metric ΔR_2 reached 147 MPa, predominantly attributable to Cottrell atmosphere formation between dislocations and mobile nitrogen interstitials.

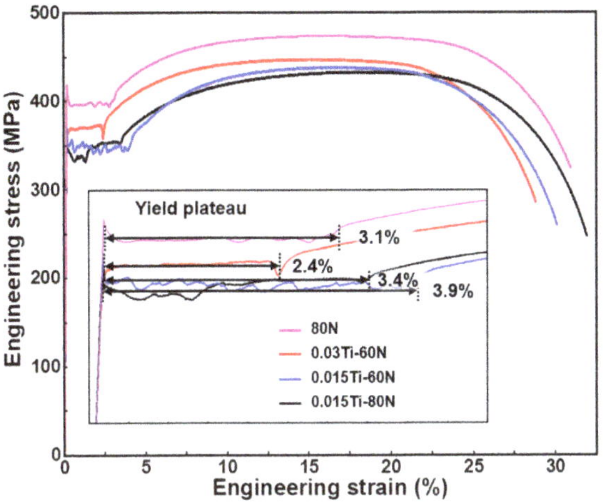

Figure 7. Engineering stress–strain curves of different experimental steels.

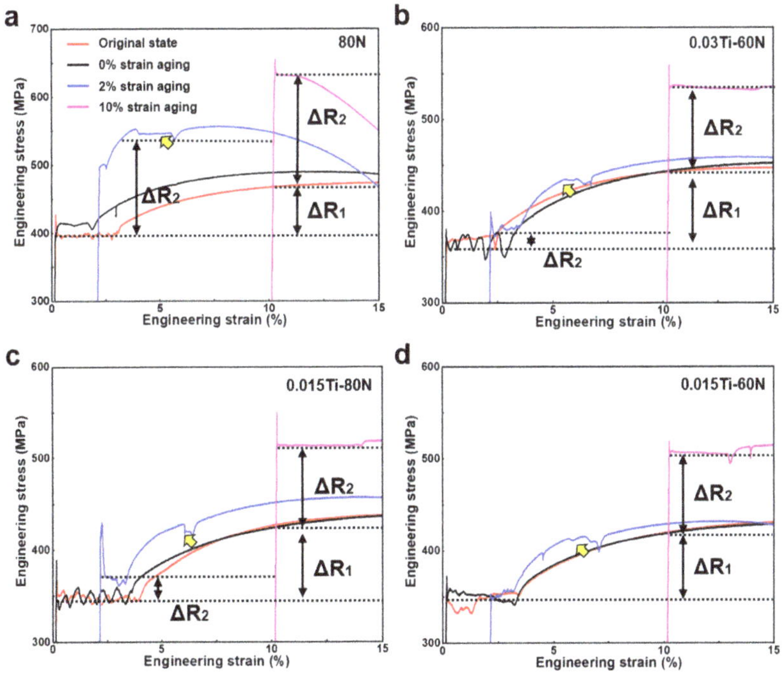

Figure 8. Engineering stress–strain curves of the experimental steels are presented as follows: Original curves and curves subjected to different strain aging conditions of 0% strain, 2% strain, and 10% strain for (**a**) 80N, (**b**) 0.03Ti-60N, (**c**) 0.015Ti-80N, and (**d**) 0.015Ti-60N, respectively. Yellow arrows indicate the positions of the secondary yielding points.

Titanium microalloying fundamentally altered this response. At identical thermomechanical processing, as shown in Table 7, the 0.03Ti-60N and 0.015Ti-80N exhibited attenuated ΔR_2 values near 30 MPa while 0.015Ti-60N demonstrated negligible ΔR_2 ($\approx$0 MPa). This progressive suppression of strain aging-induced hardening correlates with Ti's efficacy in immobilizing N through precipitate formation and solute trapping. Although residual two-stage yielding persisted in Ti-modified steels, its diminished intensity reflects substantially reduced Cottrell atmosphere density. The 0.015Ti-60N composition, achieving near-complete strain aging resistance, validates stoichiometric optimization for interstitial control in scrap-based SAPH440 production.

Table 7. Strength increments from work hardening (ΔR1) and strain aging (ΔR2) of the experimental steels under different pre-strain aging conditions.

Sample	ΔR_1/MPa (2%)	ΔR_2/MPa (2%)	ΔR_1/MPa (10%)	ΔR_2/MPa (10%)
80N	-	147 ± 2.5	78 ± 6.8	164 ± 7.2
0.03Ti-60N	-	35 ± 3.7	86 ± 7.1	88 ± 3.4
0.015Ti-80N	-	28 ± 2.1	88 ± 5.9	84 ± 2.8
0.015Ti-60N	-	0	89 ± 3.6	74 ± 4.3

Under 10% pre-strain and subsequent aging, all compositions exhibited a singular, continuous yield plateau (Figure 8), contrasting sharply with the serrated or multi-stage yielding observed at lower strains. This mechanical homogenization stems from extensive dislocation multiplication during severe plastic deformation, which simultaneously induces pronounced work hardening and fundamentally alters interstitial–dislocation interactions. Consequently, all steels manifested substantial strength elevation coupled with severe ductility reduction—consistent manifestations of pervasive strain aging. The underlying mechanism involves dislocation-mediated interstitial redistribution. During plastic deformation, crystalline defects propagate through the body-centered cubic lattice, generating high-density dislocation networks. Mobile N atoms rapidly diffuse toward tensile stress fields surrounding these dislocations, forming ordered Cottrell atmospheres that effectively pin dislocation motion. This dynamic explains the direct correlation between residual N content and strain aging intensity while higher N availability accelerates atmosphere formation, amplifying strength increases while degrading ductility. The Ti microalloyed steels demonstrated markedly attenuated property shifts despite identical thermomechanical processing. This suppression effect, quantitatively validated through internal friction spectroscopy, operates through Ti's dual interstitial management mechanisms. Thermodynamic stabilization via TiN precipitation reduces free nitrogen concentration, and substitutional solute-dislocation interactions impede interstitial mobility. The resultant significant reduction in strain aging induced hardening confirms Ti's efficacy in disrupting Cottrell atmosphere formation kinetics. Particularly in optimized Ti/N stoichiometries (0.015Ti-60N), near-complete mitigation of aging-induced embrittlement was achieved while maintaining SAPH440 mechanical specifications, which is a critical advancement for scrap-based structural steels requiring cold formability.

Impact energy measurements were conducted on the steels both before and after strain aging to assess the influence of changes in N occurrence states after strain aging (Table 8). Based on the impact energy, the strain-aging sensitivity coefficient (C_V) was calculated by the following:

$$C_V = (A_k - A_{ks})/A_k \tag{3}$$

where A_k represents the average impact absorption energy of the experimental steel before strain aging, and A_{ks} is the average impact absorption energy after the specified strain aging treatment, as per GB/T4160−2004 [24]. As shown in Figure 9, the addition of Ti significantly reduces the strain aging sensitivity coefficient of the experimental steels. For the 80 ppm N level, the C_V value reaches 7.43% for the 2% strain-aged sample. However, when Ti is added, the sensitivity coefficient decreases by 50%. Additionally, when comparing the C_V values of 15Ti-60N and 30Ti-60N, it is evident that as the Ti content increases, the strain aging sensitivity coefficient continues to decrease. This trend aligns with the internal friction test results, which indicate that Ti effectively stabilizes the free N atoms. These findings demonstrate the effective improvement of strain aging resistance in Ti-containing experimental steels. The addition of Ti helps reduce the strain aging sensitivity, with the sensitivity coefficient decreasing significantly as Ti content increases. This result highlights the critical role of Ti in enhancing the steel's performance under strain aging conditions.

Table 8. Impact energy and strain aging sensitivity coefficient of experimental steels before and after different strain aging conditions.

Sample	A_k/J	A_{ks}/J (2%)	A_{ks}/J (10%)	C_V (2%)	C_V (10%)
80N	30.29	28.04	25.82	7.43	14.76
0.03Ti-60N	32.0	31.71	30.57	0.91	4.47
0.015Ti-80N	31.42	30.14	27.76	4.07	11.65
0.015Ti-60N	33.15	32.57	31.14	1.75	6.06

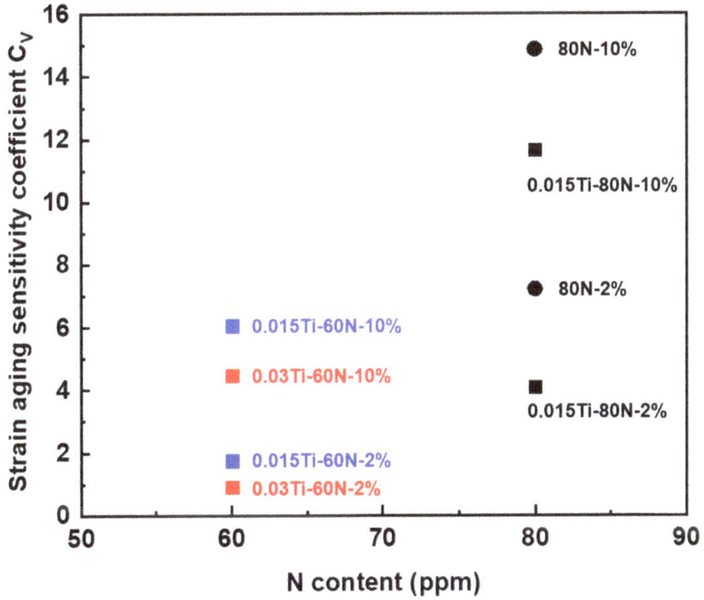

Figure 9. Effect of N and Ti on strain aging sensitivity coefficient C_V of experimental steels.

4. Conclusions

In summary, the present work investigates the influence of Ti microalloying on the N occurrence states and mechanical properties of experimental steels produced via the scrap-EAF process. The established TiN precipitation criterion based on the solubility product effectively guides Ti addition to suppress coarse liquid TiN inclusions while promoting fine solid-state TiN precipitates. Ti addition reduces free N atoms by 60–70% through efficient trapping mechanisms, which prevents the formation of Cottrell atmospheres and consequently leads to a dramatic reduction in strain aging sensitivity, with the strain aging index (ΔR_2) decreasing from 147 MPa to nearly 0 MPa. This work presents an effective strategy for controlling nitrogen-induced strain aging in scrap-based EAF steels through tailored Ti microalloying, supported by thermodynamic calculations and experimental validation of both precipitation criteria and interstitial trapping mechanisms.

Author Contributions: Conceptualization, Y.H., H.Y.; methodology, Y.H., H.Y. and C.Z.; validation, Y.H. and H.Y.; formal analysis, Y.H. and H.Y.; investigation, Y.H., H.Y., J.G., H.Z. and H.W.; resources, C.Z. and X.M.; data curation, Y.H., H.Y., J.L., S.W. and J.G.; writing—original draft preparation, Y.H. and H.Y.; writing—review and editing, C.Z., S.W., Y.H., H.Y.; visualization, Y.H., B.P., J.W. and H.Y.; supervision, S.W., X.M.; project administration, S.W., X.M.; funding acquisition. All authors have read and agreed to the published version of the manuscript.

Funding: This research was funded by the National Science and Technology Major Project (No. 2025ZD0611202), the National Key R&D Program of China (No. 2021YFB3702403), the National Natural Science Foundation of China (Nos: 52371017, 52293395, 52401137, 11290234) and the open Fund project of State Key Laboratory of Materials Processing and Die & Mould Technology (P2024-005).

Data Availability Statement: The original contributions presented in this study are included in the article. Further inquiries can be directed to the corresponding author.

Conflicts of Interest: The authors state that they have no known competing financial interests or personal relationships that could have appeared to influence the work reported in this paper. Author J.W., B.P. was employed by the company Hunan Iron and Steel Group Co., Ltd. The remaining authors declare that the research was conducted in the absence of any commercial or financial relationships that could be construed as a potential conflict of interest.

References

1. World Steel Association. *Climate Change and the Production of Iron and Steel*; World Steel Association: Brussels, Belgium, 2021.
2. Hasanbeigi, A. *Steel Climate Impact: An International Benchmarking of Energy and CO_2 Intensities*; Global Efficiency Intelligence: St. Petersburg, FL, USA, 2022.
3. Zhang, F.; Li, J.; Liu, W.; Jiao, A. The Thermodynamics and Kinetics of a Nitrogen Reaction in an Electric Arc Furnace Smelting Process. *Materials* **2023**, *16*, 33. [CrossRef] [PubMed]
4. Mo, Z.; Chu, X.; Fang, Y.; Tang, X.; Yu, X.; Zhao, Z. Effect of Nitrogen on the Microstructure and Strengthening Mechanism of High-Performance Tinplate. *J. Mater. Res. Technol.* **2024**, *33*, 4685–4698. [CrossRef]
5. Simmons, J.W. Overview: High-Nitrogen Alloying of Stainless Steels. *Mater. Sci. Eng. A* **1996**, *207*, 159–169. [CrossRef]
6. Su, Y.; Zhang, S.; Jiao, S.; Shi, X.; Yan, W.; Rong, L. Nitrogen Enhances Microstructural Thermal Stability of Si-Modified Fe-Cr-Ni Austenitic Stainless Steel. *J. Mater. Sci. Technol.* **2025**, *226*, 270–289. [CrossRef]
7. Zhao, F.; Liu, X.; Zhang, Z.; Xie, J. Effect of Nitrogen Content on the Mechanical Properties and Deformation Behaviors of Ferritic-Pearlitic Steels. *Mater. Sci. Eng. A* **2022**, *855*, 143918. [CrossRef]
8. Hwang, B.C.; Lee, T.H.; Park, S.J.; Oh, C.S.; Kim, S.J. Ductile-to-Brittle Transition Behavior of High-Nitrogen 18Cr-10Mn-0.35N Austenitic Steels Containing Ni and Cu. In *PRICM7*; Trans Tech Publications Ltd.: Baech, Switzerland, 2010; pp. 158–161. [CrossRef]
9. Xu, Z.; Wang, P.; Shen, Y.; Zhang, P.; Wang, B.; Shao, Z.; Zhang, Z. Modeling the Effect of TiN Inclusion on Fatigue Lives of High-Strength Steel. *Int. J. Fatigue* **2025**, *200*, 109090. [CrossRef]
10. Yang, R.; Li, Y.; Sun, M.; Ma, S.; Mao, Y.; Ma, Y.; Li, T.; Wang, Y.; Nie, D.; Jiang, Z. Basic Research on the Effect of Nitrogen Content on Nitride Formation and Grain Size in Titanium Microalloyed Low Carbon Steel. *J. Mater. Res. Technol.* **2025**, *36*, 337–349. [CrossRef]
11. Neuschütz, D.; Spirin, D. Nitrogen Removal and Arc Voltage Increase in EAF Steelmaking by Methane Injection into the Arc. *Steel Res. Int.* **2003**, *74*, 19–25. [CrossRef]
12. Erwee, M.W.; Pistorius, P.C. Nitrogen in SL/RN Direct Reduced Iron: Origin and Effect on Nitrogen Control in EAF Steelmaking. *Ironmak. Steelmak.* **2012**, *39*, 336–341. [CrossRef]
13. Pal, J. Thermodynamic Analysis of Nitrogen Removal in EAF by DRI Fines Injection. *Ironmak. Steelmak.* **2006**, *33*, 465–470. [CrossRef]
14. Zhou, J.; Kang, Y.; Mao, X. Precipitation Characteristic of High Strength Steels Microalloyed with Titanium Produced by Compact Strip Production. *J. Univ. Sci. Technol. Beijing Miner. Metall. Mater.* **2008**, *15*, 389–395. [CrossRef]

15. Kunze, J.; Beyer, B.; Oswald, S.; Gruner, W. Thermodynamic Data of the Formation of Titanium Nitride in Iron. *Steel Res.* **1995**, *66*, 161–166. [CrossRef]
16. Kunze, J. Solubility of Titanium Nitride in Delta Iron. *Steel Res.* **1991**, *62*, 430–432. [CrossRef]
17. Wang, S.-C. The Effect of Titanium and Nitrogen Contents on the Microstructure and Mechanical Properties of Plain Carbon Steels. *Mater. Sci. Eng. A* **1991**, *145*, 87–94. [CrossRef]
18. Du, L.; Li, W.; Zhao, S.; Wang, S. Internal Friction Study of Aging Hardening and Kinetics in Low Carbon Steel. *Procedia Eng.* **2017**, *207*, 645–650. [CrossRef]
19. Leap, M.J.; Wingert, J.C. The Effects of Grain-Refining Precipitates on the Development of Toughness in 4340 Steel. *Metall. Mater. Trans. A* **1999**, *30*, 93–114. [CrossRef]
20. Shu, Q.; Ville, V.; Tuomas, A.; Timo, F. Model for inclusion precipitation kinetics during solidification of steel applications in MnS and TiN inclusions. *Metall. Mater. Trans. B* **2020**, *51*, 2905–2916. [CrossRef]
21. Li, B.; Shi, X.; Guo, H.; Guo, J. Study on Precipitation and Growth of TiN in GCr15 Bearing Steel. *Materials* **2019**, *12*, 1463. [CrossRef] [PubMed]
22. Yuan, H.; Chen, X.; Li, L.; Chen, X.; Zhong, H.; Zhai, Q. Precipitation behavior of TiN during the solidification of high-titanium steel. *Metall. Mater. Trans. B* **2024**, *55*, 4. [CrossRef]
23. Gui, L.; Chen, D.; Liu, P.; Yu, H.; Liu, T.; Long, M.; Duan, H.; Cao, J. Precipitation of Ti-C-N Particles in Austenite During Cooling Process of High-Ti Microalloyed Steel. In Proceedings of the Materials Science & Technology 2017 (MS&T17), Pittsburgh, PA, USA, 8–12 October 2017.
24. *GB/T 4160-2004*; Steel—Test Method for Strain Ageing Sensitivity—Charpy Impact Method. State Administration for Market Regulation (Standardization Administration of the People's Republic of China): Beijing, China, 2004.

Disclaimer/Publisher's Note: The statements, opinions and data contained in all publications are solely those of the individual author(s) and contributor(s) and not of MDPI and/or the editor(s). MDPI and/or the editor(s) disclaim responsibility for any injury to people or property resulting from any ideas, methods, instructions or products referred to in the content.

materials

MDPI

Article

Rapid Stress Relief of Ti-6Al-4V Titanium Alloy by Electropulsing Treatment

Aprilia Aprilia [1], Jin Lee Tan [1,2], Zixuan Ling [1,2], Vincent Gill [3], Paul Williams [4], Martyn A. Jones [4] and Wei Zhou [1,2,*]

1 Rolls-Royce@NTU Corporate Laboratory, Nanyang Technological University, 65 Nanyang Drive, Singapore 637460, Singapore
2 School of Mechanical and Aerospace Engineering, Nanyang Technological University, 50 Nanyang Avenue, Singapore 639798, Singapore
3 Rolls-Royce Singapore Pte. Ltd., 1 Seletar Aerospace Crescent, Singapore 797565, Singapore
4 Rolls-Royce PLC, P.O. Box 31, Derby DE24 8BJ, UK
* Correspondence: mwzhou@ntu.edu.sg or wzhou@cantab.net; Tel.: +65-6790-4700

Abstract

This study investigates the effectiveness and underlying mechanisms of electropulsing treatment (EPT) for rapid stress relief of Ti-6Al-4V titanium alloy. Stress relief is an essential step in manufacturing processes to ensure long component lifespan. Residual stress accumulation within a component is often undesirable, as it may lead to premature failures. Currently, the stress relief of titanium alloys is typically carried out using an annealing heat-treatment process in a vacuum furnace. However, this method is time-consuming, usually requiring several hours. In this paper, an alternative fast stress relief method by EPT was investigated. A controllable pulsing treatment using alternating high density pulsing current with short pulse width was carried out. Results showed that EPT is effective in relieving residual stress in Ti-6Al-4V alloy. Up to 90% of the surface residual stresses induced by shot peening were successfully relieved by EPT with a treatment duration of only 114 ms. Reductions of low-angle grain boundaries (2–10°), local misorientation, and deformed grains were observed, while no significant grain growth or phase transformation was found. The stress-relief mechanism of EPT is attributed to the combined effects of dislocation movement driven by electron wind force (EWF), dislocation creep at elevated temperatures, and dislocation glide due to local yielding of residual stress under high-temperature conditions. The temperature rise during EPT was identified as a significant factor enabling stress relaxation.

Keywords: electropulsing treatment; rapid stress relief; residual stress; Ti-6Al-4V

1. Introduction

Titanium alloys are commonly used in aerospace applications because of their desirable properties, such as a high strength-to-weight ratio and excellent corrosion resistance [1]. During both the manufacturing and repair of aerospace titanium components, residual stresses are often introduced due to plastic deformation and thermal gradients generated by these processes [2]. Residual stress, especially tensile residual stress, is undesirable, as it can contribute to premature component failure [3]. Therefore, a stress-relief process is typically required to ensure adequate component lifespan.

Stress relief of titanium alloys is typically carried out through an annealing heat treatment in a vacuum furnace. However, this process is time-consuming and often requires several hours. Eshawish et al. [4] performed stress-relief heat treatment on selective laser melting-fabricated Ti-6Al-4V samples at 704 ± 14 °C for 2 ± 0.25 h in vacuum. Similarly, Emanuelli et al. [5] conducted stress-relief heat treatment inside a vacuum furnace for their laser powder bed fusion Ti-6Al-4V samples, applying a treatment temperature of 600 °C for 4 h. Kovalchuk et al. [6] also carried out stress-relief annealing of Ti-6Al-4V samples produced by wire-electron beam melting, using a treatment temperature of 650 °C for 4 h in a vacuum. Xie et al. [7] evaluated the effectiveness of stress relief in welded Ti-6Al-4V samples heat-treated at two temperatures (500 °C and 650 °C) for various soaking durations (0, 10, 30, 60, and 120 min). They found that the degree of stress relief increased with both temperature and soaking time. Under their optimal heat-treatment condition, 650 °C for 120 min, the residual stress was reduced from 550 MPa (as-welded condition) to 50 MPa. The required soaking duration for furnace-based stress relief is highly dependent on sample thickness [8]. According to AMS2801C, stress-relief furnace heat treatment for titanium alloy components should be performed at 593 °C with a soaking duration of 2 h ± 15 min [9].

In this study, a rapid stress-relief method using a controllable electropulsing treatment (EPT) is proposed. Unlike conventional heat treatment, where heating results from the transfer of thermal energy from the exterior to the interior of the material, heating in EPT is generated internally through resistive Joule heating induced by the applied current. Due to this difference in heating mechanism, EPT can achieve uniform through-thickness heating with a significantly higher heating rate, depending on the input current.

Several studies have demonstrated the use of EPT for residual stress relief. Pan et al. [10] observed a reduction in residual stress in ISO 5832-7 high-elastic cobalt-based alloy after high-density electropulsing treatment using a capacitor discharge pulse waveform with a peak current of 650 A and a pulse duration of 500 μs. The stresses were reduced from 485 MPa to 271 MPa. Similarly, Haque et al. [11] reported reductions in residual stress in a welded joint of 316L stainless steel using low-frequency direct-current (DC) electropulsing; compressive stresses were reduced from −493 MPa to −308 MPa. Xiang et al. [12] also demonstrated stress relief in quenched X80 steel samples treated with electropulsing at various current densities (3942–5562 A/mm^2). They found that stress relief increases with increasing current density, with approximately 80% reduction achieved at 5562 A/mm^2. They concluded that electropulsing accelerates dislocation annihilation by drift electrons by influencing the thermally activated dislocation motion in addition to exerting electron wind forces to the dislocation. Gudur et al. [13] applied in situ electropulsing during wire arc additive manufacturing to reduce residual stresses in the built component. DC pulses with a peak current of 800 A were applied to each deposited layer, resulting in a 24.0–29.4% reduction in residual stress compared with untreated samples. A study from our research group by Bhowmik et al. [14] also showed that EPT can promote dislocation disentanglement in the cold-worked layer of shot-peened AISI 1020 steel.

Some studies have explored the use of EPT on Ti-6Al-4V; however, these efforts focused on enhancing the alloy's plasticity and formability. Shivaprasad et al. [15] reported a reduction in V-bending force in Ti-6Al-4V subjected to pulsed direct-current electropulsing. Subrahmanyam et al. [16] investigated the effect of electric pulse parameters on the draw-bending formability of Ti-6Al-4V and found that increasing current density reduced the required forming force even at constant energy density. Li et al. [17] showed improvements in bending formability and tensile elongation with electropulsing, successfully eliminating cracking during bending. Similarly, Ao et al. [18] reported enhanced tensile elongation after electropulsing, attributing it to recrystallization of Ti-6Al-4V. They suggested that electropulsing accelerates phase transformation and recrystallization at relatively low temperatures compared with conventional heat treatment.

To date, no study has investigated the use of EPT for stress relief of Ti-6Al-4V titanium alloy. This represents a significant gap, as effective rapid stress-relief techniques would be highly valuable for aerospace manufacturing and repair applications. In this work, the effectiveness and mechanisms of EPT for rapid stress relief in Ti-6Al-4V alloy are investigated. The effect of Joule heating-induced temperature rise during EPT on the residual stress relaxation is examined. EPT was conducted using a bespoke system capable of supplying a controllable current waveform to the material. Various alternating-current cycles of high current density and short pulse duration were tested. For the pre-treatment stressed coupon, shot-peened coupons were utilized. Although the compressive residual stresses generated by shot peening are typically beneficial, this method was selected because it produces consistent and uniform surface residual stresses in the samples. This would make it easier to study the stress relief effectiveness of different EPT parameters and to evaluate the microstructure changes before and after EPT. However, it is to be noted that the intended application of EPT is not to relieve stresses induced by shot peening.

2. Materials and Methods

2.1. Shot-Peened Ti-6Al-4V Plate

Annealed Ti-6Al-4V plates with dimensions of 100 mm × 20 mm × 3.5 mm were shot-peened. The annealed Ti-6Al-4V sample has the typical annealed Ti-6Al-4V microstructure, consisting of equiaxed α grains with intergranular β phases. Shot peening was performed using cast steel shots ASR 230 at an Almen intensity of 0.097 mmA with 100% coverage. ASR 230 refers to aerospace regular-hardness shots (45–52 HRC) with a nominal size of 0.023 inches, corresponding to a diameter of approximately 0.6–0.7 mm.

Figure 1 illustrates the shot peening setup and wire-cutting dimensions. During peening, the annealed Ti-6Al-4V plate was secured in a fixture using four screws. After peening, the plate was cut into smaller coupons (50 mm × 5 mm × 3.5 mm) for electropulsing treatment (see Figure 1c). Residual stress measurements were conducted before and after EPT.

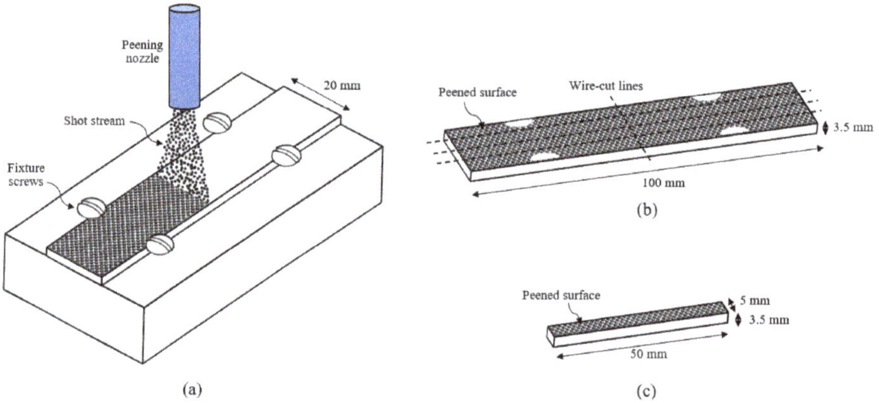

Figure 1. (**a**) Shot peening setup illustration, (**b**) wire-cutting dimensions of shot-peened samples, and (c) EPT samples' dimensions.

2.2. Residual Stress Measurement

A Stresstech G3R X-ray diffractometer was used to measure the residual stress. The modified Chi mode method, as described in ASTM E2860-20 [19], was employed, together with a Cu-Kα X-ray source and a 2 mm collimator. Measurements were conducted at a tube voltage of 30 kV and a tube current of 10 mA, with an exposure time of 100 s. The diffraction angle (2θ) was measured at 140° for the (213) crystallographic plane. Prior to measurement, the equipment was calibrated using a stress-free powder and a standard stress reference specimen. Residual stress was measured at three different locations for each sample condition. Figure 2 shows the residual stress measurement setup and the measurement locations on the samples.

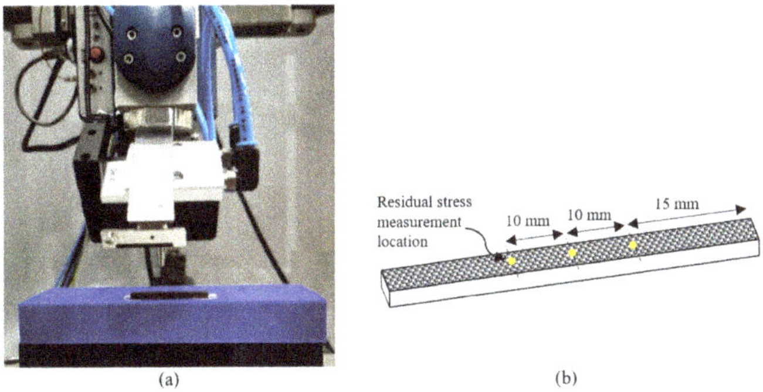

Figure 2. Residual stress measurement: (**a**) measurement setup and (**b**) measurement locations on the samples.

2.3. *Electropulsing Treatment (EPT)*

EPT was performed using a bespoke system capable of generating a controllable alternating-current (AC) square waveform. AC current was selected because the periodic reversal of electron flow is expected to enhance stress-relief effectiveness. Figure 3 shows the sample clamping setup and the alternating-current waveform used. The sample was placed on top of two copper electrodes and secured with two screws (Figure 3a), with the shot-peened surface facing upward. The sample was mechanically constrained by the screws throughout the treatment. During a localized or non-localized electropulsing treatment, the treated region is always constrained, either by the surrounding material or, as in this study, by the electrode clamping points. A square AC waveform was then supplied at a constant peak current for a specific number of cycles (Figure 3b). Each cycle consists of one positive and one negative current pulse. The duration of peak current in each pulse is referred to as the pulse width.

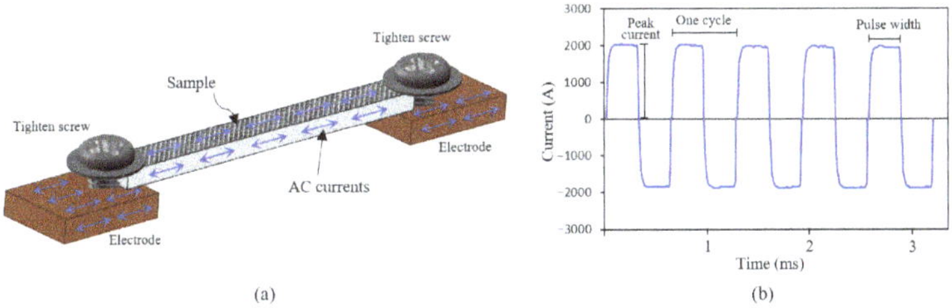

Figure 3. Electropulsing treatment: (**a**) clamping setup and (**b**) square alternating current waveform.

Five electropulsed samples were used for the investigation. The detailed parameters are shown in Table 1. The peak current and pulse width were fixed at 2000 A and 3 ms, respectively. These high current and short pulse width parameters were chosen to maximize electron–material interaction. The number of cycles varied from 10 to 31 cycles. An increase in cycles will increase the energy input and treatment duration, thereby resulting in higher sample temperatures. Figure 4 shows a photograph taken during the EPT of sample EPT5, in which the sample glowed bright red during pulsing. To record the peak temperature, a K-type thermocouple was attached to each sample during the EPT process. The peak temperatures recorded for all samples are presented in Table 1.

Table 1. EPT parameters of the five investigated samples along with the measured peak temperatures by thermocouple.

Sample	Peak Current (A)	Pulse Width (ms)	Number of Cycles	Thermocouple Measured Peak Temperature (°C)
EPT1	2000	3	10	333
EPT2	2000	3	13	475
EPT3	2000	3	19	636
EPT4	2000	3	23	741
EPT5	2000	3	31	877

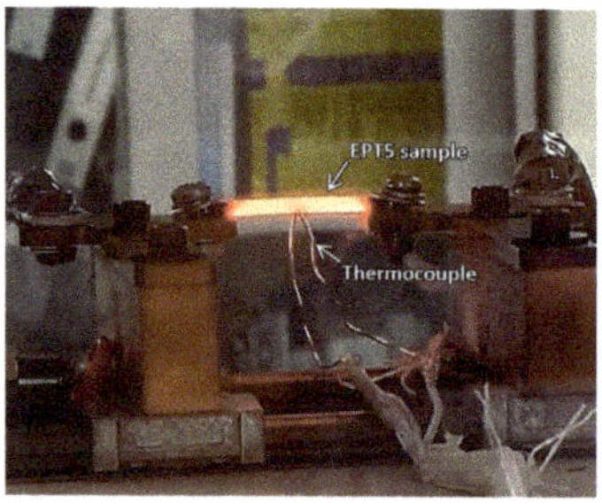

Figure 4. Photograph of the EPT5 sample during electropulsing.

2.4. Microstructure Characterization

Microstructures of as-received and as-peened samples were examined using a JEOL JSM-7600F field emission scanning electron microscope (FESEM) (Tokyo, Japan). The samples were ground with sandpaper of up to 4000 grit and subsequently polished using a 1 μm diamond suspension followed by a mixture of OP-S Struers colloidal silica suspensions and 19% hydrogen peroxide. They were then etched with Kroll's reagent (2–6% nitric acid, 1–3% hydrofluoric acid and 91–97% water) obtained from Best Chemical Co (S) Pte. Ltd. (Singapore, Singapore) for approximately 10 s.

For backscattered electron (BSE) imaging, the samples underwent the same preparation steps except for the etching process. Etching was omitted to avoid inaccuracies in β-phase size measurement that may arise from variations in etchant–material interactions across different samples. BSE micrographs were obtained at an accelerating voltage of 20 kV using the same FESEM system. β-phase size measurement was performed using the ImageJ 1.53 e (NIH, Bethesda, MD, USA).

For the electron backscattered diffraction (EBSD) analysis, EBSD maps were acquired using the same FESEM equipped with an Oxford Instruments detector (Abingdon, UK). A 20 kV accelerating voltage and a 0.1 μm step size were used. Grain boundary maps, kernel average misorientation (KAM) maps, and recrystallization fraction maps were generated.

3. Results

3.1. Residual Stress Results

Table 2 presents the residual stress values before and after electropulsing treatment, as measured using X-ray diffraction method described in Section 2.2. The residual stresses induced by shot peening were approximately −520 MPa (compressive) in both the longitudinal and transverse directions, and these values were consistent across all samples. After electropulsing, residual stress reduced in every sample. For sample EPT1, the residual stresses decreased by 23% (longitudinal) and 29% (transverse). In sample EPT2, the reductions were 53% (longitudinal) and 61% (transverse). Sample EPT3 exhibited reductions of 90% (longitudinal) and 91% (transverse). For samples EPT4 and EPT5, the residual stresses were almost completely relieved. These results show that increasing the number of electropulsing cycles enhances the degree of residual stress reduction. In this experiment, sample EPT4, treated with 23 cycles of 2000 A alternating current and a pulse width of 3 ms, achieved nearly complete relief of the initial compressive residual stress (−520 MPa), corresponding to more than a 97% reduction.

Table 2. Residual stresses (RS) before and after EPT.

Sample	Pre-EPT RS (MPa)		Post-EPT RS (MPa)		RS Reduction [a]	
	Longitudinal	Transverse	Longitudinal	Transverse	Longitudinal	Transverse
EPT1	−521 ± 17	−591 ± 26	−401 ± 8	−421 ± 7	23%	29%
EPT2	−506 ± 26	−530 ± 17	−238 ± 39	−205 ± 15	53%	61%
EPT3	−489 ± 1	−502 ± 2	−47 ± 8	−43 ± 5	90%	91%
EPT4	−513 ± 9	−523 ± 11	6 ± 7	−16 ± 6	99% *	97%
EPT5	−513 ± 6	−527 ± 30	40 ± 6	13 ± 4	92% *	98% *

Residual stress sign indication: positive value → tensile; negative value → compressive. [a] Percentage reductions are calculated using absolute value. * Residual stress has changed from compressive to tensile.

3.2. Microstructure Results

Figure 5 shows the SEM micrographs of the samples before and after shot peening (prior to EPT). As observed in the figure, the shot-peened sample exhibits a convex, curved surface. This curvature results from the repeated impact of the shots, which impart sufficient force to induce plastic deformation at the surface. The microstructure in the bulk region of the shot-peened sample remains similar to that of the as-received annealed plate, consisting of equiaxed α grains with intergranular β phases.

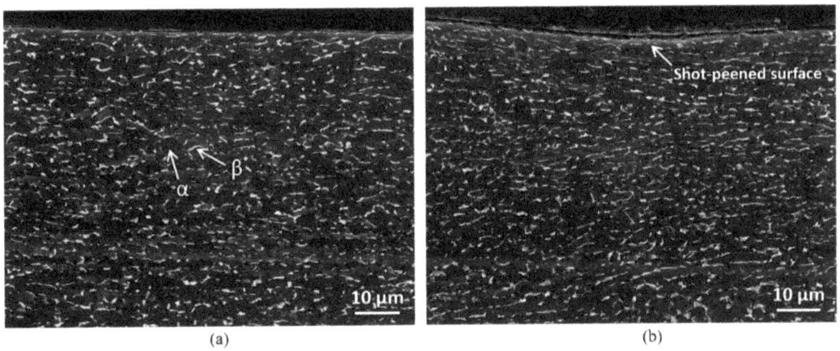

Figure 5. SEM micrographs of the sample: (**a**) before shot peening (**b**) after shot peening.

In the EPT-treated samples, the curvature of the shot-peened surface remains visible; however, some microstructure changes are observed in the bulk region. Figure 6 presents the bulk microstructures of the as-peened sample and the EPT-treated samples (EPT1, EPT2, EPT3, EPT4, and EPT5). As shown, the microstructure of sample EPT5 has changed significantly after treatment, exhibiting lamellar α colonies within prior β grains. This type of microstructure is typically observed in Ti-6Al-4V alloy when it is heated above the β-transus temperature [20]. For the remaining samples, the microstructures consist of equiaxed α grains with intergranular β phases, indicting that these samples were heated below the β-transus temperature. However, among these samples, the β-phase size in EPT4 appears to have increased.

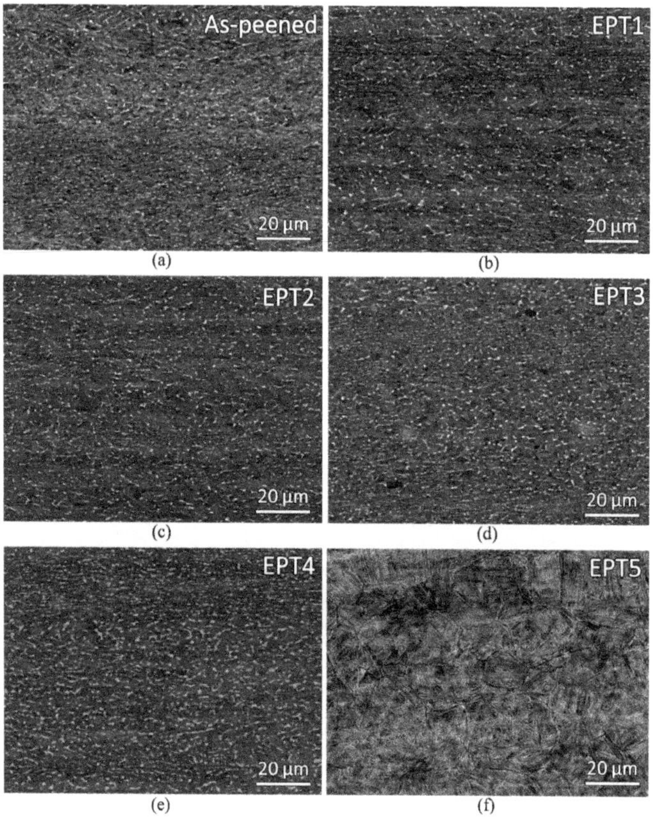

Figure 6. Backscattered electron (BSE) micrographs of the as-peened and electropulsing treated samples: (**a**) as-peened, (**b**) EPT1, (**c**) EPT2, (**d**) EPT3, (**e**) EPT4, and (**f**) EPT5.

Table 3 summarizes the observed microstructures and the average β-phase size. The β-phase in sample EPT4 is approximately 30% larger than that of the as-peened sample, based on measurements obtained using ImageJ software. This indicates that although the EPT applied to sample EPT4 (23 pulses) did not heat the material above the β-transus temperature, it still raised the temperature sufficiently to cause noticeable microstructural changes, particularly β-phase growth. To avoid significant microstructure alteration, EPT of Ti-6Al-4V alloy using a 2000 A alternating current with a 3 ms pulse width should be limited to fewer than 23 cycles. In this study, sample EPT3 satisfies this requirement, exhibiting no noticeable microstructural changes while still achieving a high residual stress reduction of approximately 90%.

Further characterization of the shot-peened region was carried out using EBSD. Figure 7 presents the grain boundary, kernel average misorientation (KAM), and recrystallization fraction maps of the as-peened sample and sample EPT3 (~90% residual stress reduction). In the grain boundary maps, red contours represent the low-angle grain boundaries (LAGBs, 2–10°), while black contours indicate high-angle grain boundaries (HAGBs, >10°). In the KAM maps, green or yellow regions indicate high local misorientation, whereas blue regions indicate low local misorientation. In the recrystallisation fraction maps, red corresponds to deformed grains, yellow to substructured grains, and blue to fully recrystallized grains. As shown in Figure 7, the amount of LAGBs in the shot-peened region decreases after EPT. Likewise, both local misorientation and the number of deformed grains are reduced following treatment. In the recrystallization map of the as-peened sample (Figure 7c), the deformed layer produced by shot peening is approximately 14 μm. After EPT, the number of deformed grains within this region is significantly reduced.

Table 3. Microstructures and average β-phase size of the samples.

Sample	Observed Microstructure	Average β-Phase Size (μm)
As-peened	Equiaxed α grains + intergranular β-phases	0.73
EPT1	Equiaxed α grains + intergranular β-phases	0.77
EPT2	Equiaxed α grains + intergranular β-phases	0.78
EPT3	Equiaxed α grains + intergranular β-phases	0.77
EPT4	Equiaxed α grains + intergranular β-phases	0.95
EPT5	Lamellar α colonies within prior β grains	-

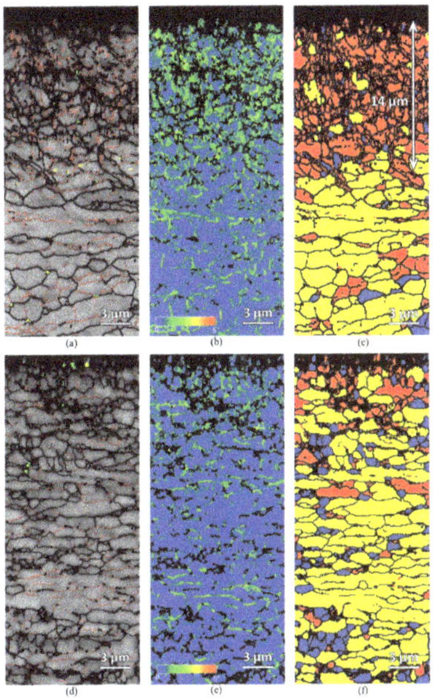

Figure 7. EBSD maps (grain boundary, KAM, and recrystallization fraction) of the as-peened sample (**a–c**) and sample EPT3 (**d–f**).

4. Discussion

4.1. Electropulsing Treatment of Ti-6Al-4V Titanium Alloy

Figure 8 summarizes the EPT results for Ti-6Al-4V titanium alloy. Electropulsing treatment with a short duration of 114 ms was able to relieve approximately 90% of the compressive residual stress without causing any observable grain growth or phase transformation. In this case, the recorded peak temperature was 636 °C. For a longer treatment duration of 138 ms, corresponding to a recorded peak temperature of 741 °C, coarsening of the β-phase grains was observed. At an even longer duration of 186 ms, with a recorded peak temperature of 877 °C, a complete change in microstructure occurred, indicating that the material had been heated above the β-transus temperature. For Ti-6Al-4V, the β-transus temperature is typically reported to be between 970 and 1000 °C [21]. These findings indicate that the peak temperatures measured by the thermocouple are not accurate. This discrepancy is understandable, as the thermocouple used had a sampling rate of only 1 Hz. Due to the extremely rapid heating associated with electropulsing, the thermocouple was unable to capture the actual peak temperature experienced by the material. To estimate the true peak temperature, a numerical simulation was performed using the ANSYS Mechanical 2020 R2 (Ansys, Canonsburg, PA, USA) with an element mesh size of 0.25 mm.

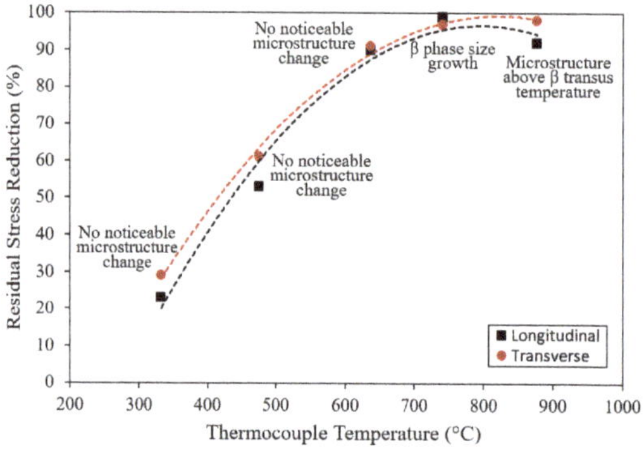

Figure 8. Summary of electropulsing treatment of Ti-6Al-4V alloy.

Table 4 presents the peak temperatures measured by the thermocouple and those computed through numerical simulation, along with the corresponding treatment durations. As shown, the simulation-predicted peak temperature for sample EPT5 exceeds the β-transus temperature of Ti-6Al-4V (970–1000 °C), confirming that the simulated values more accurately reflect the true temperatures experienced during EPT. For sample EPT3, the ideal case with no observable grain growth or phase transformation, the simulation-computed peak temperature is approximately 785 °C.

Table 4. Treatment duration and peak temperatures of the samples (measured by thermocouples and computed by simulation).

Sample	Treatment Duration (ms)	Thermocouple-Measured Peak Temperature (°C)	Simulation-Computed Peak Temperature (°C)
EPT1	60	333	451
EPT2	78	475	571
EPT3	114	636	785
EPT4	138	741	969
EPT5	186	877	1270

4.2. Understanding Stress Relief Mechanisms of Electropulsing Treatment

Stress relief in a material occurs when dislocations move, thereby releasing stored strain energy and reducing internal stress. During electropulsing treatment, a high density of electric current passes through the material, resulting in a force acting on atoms and dislocations due to their interaction with the flowing electrons. This force is known as the electron wind force (EWF) [22]. The EWF functions as a mechanical force that pushes dislocation to move in the direction of electron flow. The movement typically occurs within individual grains, as movement across grain boundaries is often impeded by the dislocation networks present at these boundaries. During the process, dislocation accumulation, reorganization, and annihilation may take place at grain boundaries. The overall stress induced by EWF can be estimated using the equation given below [22,23].

$$\sigma_{ed} = C_{ed} \cdot J, \quad (1)$$

where σ_{ed} is the stress induced by EWF, C_{ed} is the coefficient of electron–dislocation interaction, and J is the peak current density. In this work, the peak current density is 1.14×10^8 A/m^2, and C_{ed} for Ti atoms is 3.2×10^{-8} MPa/(Am^{-2}) [22]. Therefore, the calculated EWF-induced stress is 3.65 MPa.

For a dislocation to move, the applied stress must exceed its Peierls stress or, for large-scale dislocation motion, the critical resolved shear stress (CRSS) [24]. For Ti-6Al-4V titanium alloy, the CRSS at room temperature is estimated to be about 380 MPa for prismatic slip {1010}<1120> of a polycrystalline Ti-6Al-4V alloy [23]. However, Salem et al. [25] reported that the CRSS of Ti-6Al-4V single-colony samples decreases to about 40–80 MPa for the same prismatic slip at 815 °C. Similar to yield strength, CRSS decreases with increasing temperature. The calculated EWF-induced stress of 3.65 MPa from Equation (1) is significantly lower than the reported CRSS values, even at elevated temperatures. Therefore, additional mechanisms must contribute to dislocation motion and the resulting residual stress reduction during EPT.

As shown in Tables 2 and 3, increasing the number of pulsing cycles or the treatment duration enhances residual stress relief, even when the peak current density remains constant. This indicates that pulsing cycles and treatment duration also play roles in stress reduction. With a larger number of cycles of a longer pulsing duration, more thermal energy accumulates (energy coupling) in the material through Joule heating, which arises from the conversion of electrical energy into heat as current passes through a resistive material (Figure 9). As the number of cycles or treatment duration increases, the material temperature rises (Equation (2)), promoting increased dislocation motion. This behavior follows the principles of material heat capacity and Joule's law of electrical heating.

$$\Delta T = \frac{I^2 R t}{mc}, \quad (2)$$

where ΔT is the increase of temperature, I is electric current, R is material resistance, t is pulsing duration, m is the mass of the material, and c is the specific heat capacity.

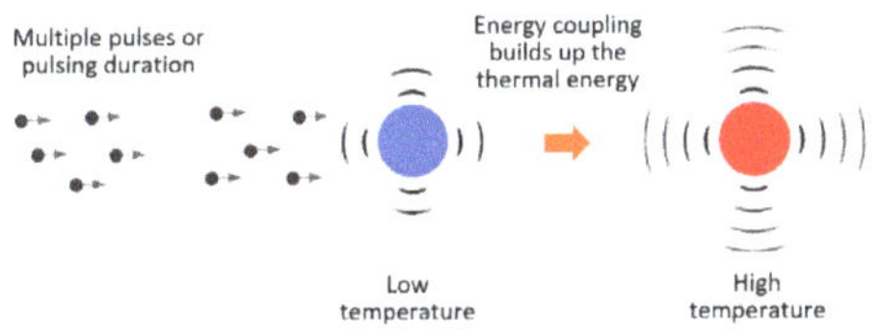

Figure 9. Joule heating in electropulsing treatment.

In addition to the overall temperature rise caused by Joule heating, there may also be a transient temperature increase during pulsing, an effect that occurs only under pulsed current conditions and not during steady direct-current heating. The transient temperature rise depends on the pulse width t_p and the thermal diffusivity α of the material [26]. For the temperature range relevant to this study, the thermal diffusivity of Ti-6Al-4V varies from 2.7 mm^2/s to 8.5 mm^2/s [27]. With a pulse width of t_p= 3 ms, the characteristic diffusion length, $l_{diff} \approx \sqrt{\alpha t_p}$, is approximately 90–160 µm. This diffusion length is much smaller than the sample thickness (3.5 mm), indicating the likelihood of localized temperature rises within the material due to the thermal flux generated as the pulsed current ramps up and down. Furthermore, localized heating may also occur at dislocation cores due to 'local' Joule heating [28]. The high electrical resistivity at dislocation cores can lead to localized heat nucleation, arising from microscopic resistivity inhomogeneities. These three phenomena, global Joule heating, transient pulsing-induced heating, and local dislocation-core Joule heating, collectively contribute to significant temperature rises at the dislocation cores during electropulsing treatment.

In summary, stress relief during electropulsing treatment can be attributed to three possible mechanisms. First, dislocation movement may be induced by the EWF, which drives dislocations in the direction of electron flow. However, as discussed earlier, the magnitude of the EWF-induced stress in this study is insufficient to produce large-scale dislocation motion on its own. Second, stress relief may occur due to the increased vacancy diffusion at elevated temperatures which allows for dislocation climb (also referred to as dislocation creep) and therefore dislocation annihilation. This process is believed to be one of the primary mechanisms for conventional thermal (static) stress relief [29]. Unlike EWF-driven motion, vacancy diffusion and thermal stress relief are non-directional and time-dependent. The increase of temperature in the material during electropulsing can result from the 'global' Joule heating, transient temperature rise from pulsing, and the local temperature rise at dislocation cores. The third possible mechanism of the stress relief is dislocation glide by local yielding of residual stresses at high temperatures. As the material temperature increases, the CRSS and yield strength of Ti-6Al-4V decrease. When the reduced yield strength falls below the existing residual stresses, plastic deformation and dislocation glides occur which results in a fall of residual stresses to the level of the yield strength.

Without further evidence, it is likely that all three mechanisms and potentially other unidentified mechanisms are interacting to result in the stress relief recorded in this study. This study shows that in using electropulsing treatment for residual stress reduction, the increase of material temperature by increasing the pulsing cycle and treatment duration is an important factor for the residual stress reduction. Unless substantially higher current densities are applied, the influence of the electron wind force is far less significant than the temperature-driven effects. In practical applications, much higher current density is not feasible due to safety concerns, large component sizes, and the substantial electrical energy required. Although electropulsing treatment involves temperature rise similar to the conventional furnace heat treatment, EPT offers several advantages, including fast heating rates, significantly shorter exposure times of the material at elevated temperatures, and the capability for localized treatment. These characteristics highlight the potential of EPT as a viable and efficient method for residual stress relief.

5. Conclusions

In this study, rapid stress relief of Ti-6Al-4V titanium alloy parts using a controllable electropulsing treatment was investigated. Electropulsing treatments with varying numbers of square alternating-current cycles, using a peak current of 2000 A and a pulse width of 3 ms, were investigated. The conclusions drawn from this study are as follows:

(1) Electropulsing treatment of Ti-6Al-4V alloy for a short period of 114 ms (19 pulse cycles) can relieve approximately 90% of the compressive residual stress without resulting in any observable grain growth or phase transformation. Further optimization of treatment parameters such as current density and pulse width can be further explored to enhance the residual stress relief.
(2) EBSD analysis showed reductions in low-angle grain boundaries (2–10°), local misorientation, and deformed grains in the electropulsed samples.

(3) Stress relief during electropulsing treatment is attributed to a combination of three possible mechanisms: EWF-induced stress, dislocation creep, and dislocation glide by local yielding. The increase of material temperature in electropulsing treatment is an important aspect of the stress relief, as it enables dislocation creep and local yielding to occur.

Author Contributions: Conceptualization, W.Z., V.G., P.W. and M.A.J.; methodology, W.Z., V.G., P.W. and M.A.J.; software, Z.L.; validation, A.A., J.L.T. and Z.L.; formal analysis, W.Z., A.A., J.L.T., V.G., P.W. and M.A.J.; investigation, W.Z., A.A., J.L.T., Z.L., V.G., P.W. and M.A.J.; resources, W.Z., V.G., P.W. and M.A.J.; data curation, A.A., J.L.T. and Z.L.; writing—original draft preparation, A.A.; writing—review and editing, W.Z., J.L.T., Z.L., V.G., P.W. and M.A.J.; visualization, A.A.; supervision, W.Z.; project administration, W.Z., V.G., P.W. and M.A.J.; funding acquisition, W.Z., V.G., P.W. and M.A.J. All authors have read and agreed to the published version of the manuscript.

Funding: This work was financially supported by National Research Foundation of Singapore, Rolls-Royce Singapore Pte. Ltd. and Nanyang Technological through grants #002123-00009 and #002124-00009.

Institutional Review Board Statement: Not applicable.

Informed Consent Statement: Not applicable.

Data Availability Statement: The original contributions presented in this study are included in the article. Further inquiries can be directed to the corresponding author.

Acknowledgments: The authors would like to thank the School of Mechanical and Aerospace Engineering at Nanyang Technological University for the research support.

Conflicts of Interest: Author Vincent Gill was employed by the company Rolls-Royce Singapore Pte Ltd. Authors Paul Williams and Martyn A. Jones were employed by the company Rolls-Royce PLC. The remaining authors declare that the research was conducted in the absence of any commercial or financial relationships that could be construed as a potential conflict of interest. The authors declare that this study received funding from Rolls-Royce Singapore Pte. Ltd. The funder was not involved in the study design, collection, analysis, interpretation of data, the writing of this article or the decision to submit it for publication.

References

1. Sala, S.T.; Keller, S.; Chupakhin, S.; Pöltl, D.; Klusemann, B.; Kashaev, N. Effect of laser peen forming process parameters on bending and surface quality of Ti-6Al-4V sheets. *J. Mater. Process. Technol.* **2022**, *305*, 117578. [CrossRef]
2. Gautam, N.; Subramaniyan, A.K.; Mondi, P.R. Evaluation methods for residual stress measurement in large components. *Mater. Today Proc.* **2021**, *44*, 4239–4244. [CrossRef]
3. Li, N.; Zhang, M.; Ye, J.L.; Liu, C. Experimental investigation on residual stress distribution in zirconium/titanium/steel tri-metal explosively welded composite plate after cutting and welding of a cover plate. *J. Manuf. Process.* **2021**, *64*, 455–463. [CrossRef]
4. Eshawish, N.; Malinov, S.; Sha, W.; Walls, P. Microstructure and mechanical properties of Ti-6Al-4V manufactured by selective laser melting after stress relieving, hot isostatic pressing treatment and post-heat treatment. *J. Mater. Eng. Perform.* **2021**, *30*, 5290–5296. [CrossRef]
5. Emanuelli, L.; Molinari, A.; Facchini, L.; Sbettega, E.; Carmignato, S.; Bandini, M.; Benedetti, M. Effect of heat treatment temperature and turning residual stresses on the plain and notch fatigue strength of Ti-6Al-4V additively manufactured via laser powder bed fusion. *Int. J. Fatigue* **2022**, *162*, 107009. [CrossRef]
6. Kovalchuk, D.; Melnyk, V.; Melnyk, I.; Savvakin, D.; Dekhtyar, O.; Stasiuk, O.; Markovsky, P. Microstructure and properties of Ti-6Al-4V articles 3D-printed with co-axial electron beam and wire technology. *J. Mater. Eng. Perform.* **2021**, *30*, 5307–5322. [CrossRef]
7. Xie, P.; Zhao, H.; Wu, B.; Gong, S. Evaluation of residual stresses relaxation by post weld heat treatment using contour method and X-ray diffraction method. *Exp. Mech.* **2015**, *55*, 1329–1337. [CrossRef]
8. Kalyankar, V.D.; Chudasama, G. Effect of post weld heat treatment on mechanical properties of pressure vessel steels. *Mater. Today Proc.* **2018**, *5*, 24675–24684. [CrossRef]
9. *AMS2801C*; Heat Treatment of Titanium Alloy Parts. SAE International: Warrendale, PA, USA, 2022.

10. Pan, L.; Wang, B.; Xu, Z. Effects of electropulsing treatment on residual stresses of high elastic cobalt-base alloy ISO 5832-7. *J. Alloys Compd.* **2019**, *792*, 994–999. [CrossRef]
11. Haque, A.; Sherbondy, J.; Warywoba, D.; Hsu, P.; Roy, S. Room-temperature stress reduction in welded joints through electropulsing. *J. Mater. Process. Technol.* **2022**, *299*, 117391. [CrossRef]
12. Xiang, S.; Zhang, X. Dislocation structure evolution under electroplastic effect. *Mater. Sci. Eng. A* **2019**, *761*, 138026. [CrossRef]
13. Gudur, S.; Simhambhatla, S.; Reddy, N.V. Residual stress reduction in wire arc additively manufactured parts using in-situ electric pulses. *Sci. Technol. Weld. Join.* **2023**, *28*, 193–199. [CrossRef]
14. Bhowmik, A.; Tan, J.L.; Yang, Y.; Aprilia, A.; Chia, N.; Williams, P.; Jones, M.; Zhou, W. Misorientation and dislocation evolution in rapid residual stress relaxation by electropulsing. *J. Mater. Sci. Technol.* **2025**, *209*, 292–299. [CrossRef]
15. Shivaprasad, C.; Subrahmanyam, A.; Reddy, N.V. Effect of electric path in electric pulse aided V-bending of Ti-6Al-4V: An experimental and numerical study. *J. Manuf. Process.* **2023**, *100*, 75–84. [CrossRef]
16. Subrahmanyam, A.; Saurabh, S.; Praveen, K.; Ramu, G.; Reddy, N.V. Electric pulse aided draw-bending of Ti-6Al-4V. In Proceedings of the Flexible Automation and Intelligent Manufacturing, Detroit, MI, USA, 19–23 June 2022. [CrossRef]
17. Li, X.; Zhou, W.; Zhao, S.; Chen, J. Effect of pulse current on bending behavior of Ti6Al4V alloy. *Procedia Eng.* **2014**, *81*, 1799–1804. [CrossRef]
18. Ao, D.; Chu, X.; Yang, Y.; Lin, S.; Gao, J. Effect of electropulsing treatment on microstructure and mechanical behavior of Ti-6Al-4V alloy sheet under argon gas protection. *Vacuum* **2018**, *148*, 230–238. [CrossRef]
19. *ASTM E2860-20*; Standard Test Method for Residual Stress Measurement by X-Ray Diffraction for Bearing Steels. ASTM International: West Conshohocken, PA, USA, 2021.
20. Pederson, R. Microstructure and Phase Transformation of Ti-6Al-4V. Ph.D. Dissertation, Luleå University of Technology, Luleå, Sweden, 2002.
21. Neikter, M.; Åkerfeldt, P.; Pederson, R.; Antti, M.L. Microstructure characterization of Ti-6Al-4V from different additive manufacturing processes. *IOP Conf. Ser. Mater. Sci. Eng.* **2017**, *258*, 012007. [CrossRef]
22. Sprecher, A.F.; Mannan, S.L.; Conrad, H. Overview no. 49: On the mechanisms for the electroplastic effect in metals. *Acta Metall.* **1986**, *34*, 1145–1162. [CrossRef]
23. Li, W.; Shen, Y.; Liu, H.; Wang, Y.; Zhu, W.; Xie, C. Non-octahedral-like dislocation glides in aluminum induced by athermal effect of electric pulse. *J. Mater. Res.* **2016**, *31*, 1193–1200. [CrossRef]
24. Hutchinson, W.B.; Barnett, M.R. Effective values of critical resolved shear stress for slip in polycrystalline magnesium and other hcp metals. *Scr. Mater.* **2010**, *63*, 737–740. [CrossRef]
25. Salem, A.A.; Semiatin, S.L. Anisotropy of the hot plastic deformation of Ti-6Al-4V single-colony samples. *Mater. Sci. Eng. A* **2009**, *508*, 114–120. [CrossRef]
26. Zhao, S.; Zhang, R.; Chong, Y.; Li, X.; Abu-Odeh, A.; Rothchild, E.; Chrzan, D.C.; Asta, M.; Morris, J.M.; Minor, A.M. Defect reconfiguration in a Ti-Al alloy via electroplasticity. *Nat. Mater.* **2021**, *20*, 468–472. [CrossRef]
27. Milošević, N.; Aleksic, I. Thermophysical properties of solid phase Ti-6Al-4V alloy over a wide temperature range. *Int. J. Mater. Res.* **2012**, *103*, 707–714. [CrossRef]
28. Zhang, X.; Li, H.; Zhan, M. Mechanism for the macro and micro behaviors of the Ni-based superalloy during electrically-assisted tension: Local joule heating effect. *J. Alloy Compd.* **2018**, *742*, 480–489. [CrossRef]
29. Kohnert, A.A.; Capolungo, L. The kinetics of static recovery by dislocation climb. *NPJ Comput. Mater.* **2022**, *8*, 104. [CrossRef]

Disclaimer/Publisher's Note: The statements, opinions and data contained in all publications are solely those of the individual author(s) and contributor(s) and not of MDPI and/or the editor(s). MDPI and/or the editor(s) disclaim responsibility for any injury to people or property resulting from any ideas, methods, instructions or products referred to in the content.

MDPI AG
Grosspeteranlage 5
4052 Basel
Switzerland
Tel.: +41 61 683 77 34

Materials Editorial Office
E-mail: materials@mdpi.com
www.mdpi.com/journal/materials

Disclaimer/Publisher's Note: The title and front matter of this reprint are at the discretion of the Guest Editors. The publisher is not responsible for their content or any associated concerns. The statements, opinions and data contained in all individual articles are solely those of the individual Editors and contributors and not of MDPI. MDPI disclaims responsibility for any injury to people or property resulting from any ideas, methods, instructions or products referred to in the content.

www.ingramcontent.com/pod-product-compliance
Lightning Source LLC
LaVergne TN
LVHW070448061126
851858LV00011B/186
* 9 7 8 3 7 2 5 8 8 2 6 3 2 *